T0212397

Compact Textbooks in Mathematics

Compact Textbooks in Mathematics

This textbook series presents concise introductions to current topics in mathematics and mainly addresses advanced undergraduates and master students. The concept is to offer small books covering subject matter equivalent to 2- or 3-hour lectures or seminars which are also suitable for self-study. The books provide students and teachers with new perspectives and novel approaches. They feature examples and exercises to illustrate key concepts and applications of the theoretical contents. The series also includes textbooks specifically speaking to the needs of students from other disciplines such as physics, computer science, engineering, life sciences, finance.

- **compact**: small books presenting the relevant knowledge
- **learning made easy**: examples and exercises illustrate the application of the contents
- **useful for lecturers**: each title can serve as basis and guideline for a semester course/lecture/seminar of 2–3 hours per week.

More information about this series at http://www.springer.com/series/11225

Benedikt Jahnel • Wolfgang König

Probabilistic Methods in Telecommunications

 Birkhäuser

Benedikt Jahnel
Weierstraß-Institut
Berlin, Germany

Wolfgang König
Technische Universität Berlin
Institut für Mathematik
Berlin, Germany

Weierstraß-Institut
Berlin, Germany

This textbook has been reviewed and accepted by the Editorial Board of Mathematik Kompakt, the German language version of this series.

ISSN 2296-4568   ISSN 2296-455X   (electronic)
Compact Textbooks in Mathematics
ISBN 978-3-030-36089-4   ISBN 978-3-030-36090-0   (eBook)
https://doi.org/10.1007/978-3-030-36090-0

Mathematics Subject Classification: 60-01, 60G55, 60D05, 60F10, 60K35, 60K37

This book is published under the imprint Birkhäuser, www.birkhauser-science.com, by the registered company Springer Nature Switzerland AG.
The registered company address is: Gewerbestrasse 11, 6330 Cham, Switzerland

Preface

This is a strongly expanded version of our notes for a lecture that we gave in the summer semester 2018 at Technische Universität Berlin. We introduced mathematical concepts and theories for the rigorous probabilistic analysis of decentralized spatial telecommunication systems, more precisely, *multihop adhoc systems*. These are spatial networks consisting of randomly located participants (users, devices, base stations), enjoying a message-relaying functionality.

The need for reliable theoretical knowledge on the functionality and properties of *wireless communication systems* becomes increasingly important in our community. New technologies based on functionalities and concepts like *device-to-device communication* (D2D), *self-driving cars*, *internet of things* (IoT), *peer-to-peer systems* (P2P) and *uberization* push the theoretical and applied research and increasingly influence our daily lives. Even though there are systems in operation that apparently work well, in order to be able to cope with the ever increasing data, it is of great importance to explore systems of new types of functionality.

One of the basic functionalities that has been under discussion for decades is the idea of a *decentralized* system, which is sustained by the devices in an adhoc manner, when each device is equipped with a relaying functionality. That is, the transmission of data is predominantly not conducted via fixed base stations but by multi-hop paths via all the devices. Such a system is likely to have a number of advantages, for example with respect to capacity, robustness against outages, expenses, and energy efficiency. Indeed, typically base stations in cellular networks are expensive and sparsely distributed. They necessitate high energy consumption for long distance transmissions, and the outage of one of them has potentially severe consequences for the whole network.

On the other hand, decentralized networks come with a number of challenges that call for deep theoretical research in various disciplines, in particular mathematics. One of the most important fields in this context is *spatial probability* or *stochastic geometry*. Indeed, even though an operator can design functionalities and certain parameters of the system, decisive elements of the network behave randomly and should therefore be described via probabilistic models. This monograph lays important foundations for that.

Probabilistic research on wireless communication is several decades old and has many facets. However, when it comes to a qualified analysis of *spatial* models for wireless communication, the history dates back only to the beginning of this millennium. The last twenty years have seen a strong adaptation and specialization of that theory towards telecommunication-oriented applications, and this development is obviously far from being complete. Stochastic geometry for telecommunication has become an interesting and useful application field, which is underlined by the appearance of first introductory monographs on this subject, [FraMee08], [BacBla09a, BacBla09b] and [Hän12]. They all address the mathematically educated engineer, computer scientist or physicist. While [Hän12] aims at comprehensiveness of point processes, connectivity, interference and coverage, [FraMee08] treats connectivity in spatial discrete and continuous network models with a special focus on information flow and navigation. Baccelli and Błaszczyszyn [BacBla09a, BacBla09b] discusses at length further-reaching topics like interference-reduction strategies and message routing.

The present text presents the cornerstones of stochastic geometry in telecommunications like random point processes (▶ Chap. 2), connectivity (▶ Chap. 4), and interference (▶ Chap. 5). However, it is written for mathematicians with a focus on probability, which allows us to cover a wider range of concepts and to reach a deeper level of probabilistic modeling. Indeed, the present text goes considerably beyond the previously mentioned monographs by treating also random environments (▶ Chap. 3), large-scale asymptotics (▶ Chap. 6), extremely rare events (▶ Chap. 7), and random propagation of malware (▶ Chap. 8). These chapters rely on recent innovative research and are presented in textbook form here for the first time.

Our goal is to present a useful wealth of theoretical material and illuminating examples. We do not give all the details of the proofs, but restrict at many places the explanations to the main idea and the proof strategy. We did not strive to give complete accounts on the theories that we employ, but restrict to those aspects that we find useful for the mathematical analysis of spatial telecommunication systems. On the other hand, we invested great care in explaining concrete examples and in listing a large number of useful exercises.

The prerequisites that we expect from our readership are a great deal of the content of two standard lectures on probability for mathematicians, notably familiarity with general measure theory, weak convergence of random variables, conditional expectations, and probability kernels. Concepts like random measures, simple versions of ergodic theorems, large deviations theory, Markov chains in continuous time and interacting particle systems will be briefly explained and further expanded for telecommunication applications.

We believe that the mathematical analysis of spatial telecommunication systems poses many far-reaching and interesting open questions. Our text is meant to provide a solid base for meeting the mathematical challenges that are to come. We hope to inspire researchers from the telecommunication community with a background in probability as well as probabilists searching for new and interesting applications, and we hope to build a bridge between them.

Berlin, Germany   Benedikt Jahnel, Wolfgang König
November 2019

Acknowledgments

We would like to thank our co-authors and co-workers Christian Hirsch, András Tóbiás and Alexander Hinsen for their contributions to the text and illustrations.

This work was supported by the WIAS Leibniz Group *Probabilistic Methods for Mobile Ad Hoc Networks* as well as the ECMath project *Data Mobility in Ad Hoc Networks: Vulnerability & Security* within the research center MATHEON. We received further funding by the German Research Foundation DFG under Germany's Excellence Strategy MATH+: The Berlin Mathematics Research Center, EXC–2046/1 project ID: 390685689. Some of the results presented in this book were initially developed within joint research projects with Elie Cali and Jean-Philippe Wary at Orange Labs Paris and funded by ORANGE S.A. We thank AIMS Ghana for their hospitality during the final phase of writing this monograph.

Contents

Introduction

In this book we present concepts, tools and methodologies for a probabilistic treatment of decentralized spatial telecommunication systems. The ubiquitous ansatz for analyzing such a device-to-device network is by means of a *random point process* with interactions and additional features. The most fundamental and flexible mathematical model is the *Poisson point process* (PPP), which has a great degree of independence and admits many explicit calculations for important expectations and functionals. We build on this fundamental structure a variety of network models intended to represent the most important aspects of the telecommunication systems. Among the additional features that we introduce are *random environments* and *interference* constraints. Our main focus is based on answering questions about the *connectivity* of the system, which we will address predominantly in *limiting settings* of a large number of network participants, both in a *typical* as well as in an *atypical* situation. Finally we will introduce models for the space-time evolution of malware in the adhoc network by means of *interacting Markov jump processes* and analyze the propagation in the presence of some countermeasures.

Here is a colloquial introduction to our basic model of a spatial telecommunication system, which will be used throughout. Let $D \subset \mathbb{R}^d$ be a domain and $\Phi = (X_i)_{i \in I}$ a random locally finite subset of D, the set of the locations of the participants, e.g., some electronic devices. Imagine that each of them has a message that should be transmitted to some location in D. Hence, the participants act as *transmitters*, and there are some locations that act as possible *receivers*. These may be other transmitters or additional locations, e.g., base stations. Each of the messages is supposed to perform a multi-hop trajectory via the transmitters, i.e., it makes a sequence of jumps from device to device until it finally arrives at its destination. All these jumps are local in the sense that their lengths do not exceed a certain threshold, $R \in (0, \infty)$. Immediately, some obvious questions arise about the success of this system, namely questions about *connectivity*:

- Does a typical message successfully reach its destination?
- How many messages successfully reach their destination?
- How do the answers depend on the locations of the transmitters and receivers, or on R, or on the placement of the cloud of all the devices?
- How do the answers change if we impose a bound on the number of hops?

© Springer Nature Switzerland AG 2020
B. Jahnel, W. König, *Probabilistic Methods in Telecommunications*, Compact Textbooks in Mathematics, https://doi.org/10.1007/978-3-030-36090-0_1

All these questions highly depend on the underlying randomness and can be answered on various levels, for example in terms of distributions or expectations.

We ask detailed questions, and we consider rather complex objects that have many details, even though the model is only a simple caricature of a real-world scenario. In order to render the model more realistic, we introduce two important additional features, *randomness in the environment* and *interference constraints* on the connections. The random environment is introduced via a random intensity landscape in D that controls the local density of the devices. This enables a much more realistic modeling, for example of an urban setting. Interference at a location in D is the joint noise that all the message transmissions create in D. As a consequence, a weak transmission attempt may then fail. This effect is ubiquitous in realistic telecommunication networks. In the light of these two extensions, the above four questions become more interesting and increase in complexity.

In our probabilistic treatment of the above questions, we will emphasize the analysis of *large systems*. We are guided by the philosophy that even small systems, which are often already pretty complex, can be well approximated by taking the limit of infinitely many devices. This approach enables us to reveal macroscopic effects and tendencies that are decisive for the performance of the network. The two main limiting scenarios that we have in mind are the following:

- *High device density:* The domain D is fixed with a very large number of participants, like a bustling square in a city or a social event where many people convene in one place.
- *Large area:* The domain D is large and the density of participants is fixed, like in everyday life in a big city.

In order to derive useful and rigorous assertions, we need to introduce and employ some comprehensive mathematical methodologies.

For a description of the large systems, we find it most important to answer the above four questions in each of the following two settings.

- *Typical behavior:* Approximation of system characteristics by averaged quantities.
- *Extreme behavior:* Estimation of the probability and analysis of the situations in which the system behaves unexpectedly.

Answers in the first setting can be used to design the system and set parameters in such a way that the system performs well on average. On the other hand, answers in the second setting address the desire of operators to understand situations in which the system is far away from the average, in particular when the network performance is extremely bad.

The situations considered so far are static, i.e., not evolving in time, and thus can be interpreted as the description of a snapshot of an adhoc system. However, we also will consider *random propagation* of messages over the system in a dynamic way. That is, we describe the time evolution of the messages of the devices. We have in mind 'unwanted' messages, i.e., malware. Furthermore, we add random mechanisms of countermeasures to the system and ask under what circumstances the malware will diminish or spread unboundedly, and in the latter case, how fast.

We come now to a more detailed description of the mathematical content of the monograph. In ▶ Chap. 2 we introduce the theory of *random point processes* in Euclidean space. We put particular weight on the *Poisson point process, Campbell moment formula, marked Poison point processes* and *Palm calculus*. The material of this chapter is standard and covered by many textbooks.

The next ▶ Chap. 3 contains less common and more original material, namely *Cox point process*, that is, Poisson point processes in *random environment*. We discuss a variety of *random tessellations* that can be used to design random street systems. We prepare for later chapters by introducing the concept of *stabilization* as a quantitative decorrelation property of the environment. Finally we present the *inversion formula for Palm calculus*, which allows to recover a distribution from its Palm version with the help of the *Voronoi tessellation*.

In ▶ Chap. 4 we start to address questions concerning *connectivity* in unbounded multi-hop systems. This lies in the realm of the celebrated field of *continuum percolation*, which asks whether or not the Poisson point process admits infinitely long multi-hop trajectories or not. The main mathematical model here is the *Boolean model*, the union of balls around the Poisson points, and the question about the sizes of the connected components. We will introduce this concept and the most important results. Additionally, we discuss percolation for Cox point processes and *hybrid systems*, where adhoc components with a bounded number of hops are used to augment an infrastructure process. ▶ Chapters 2 and 4 summarize the basics of a probabilistic theory that is sometimes called *stochastic geometry* and is the focus of many specialized lectures on probability because of its universal value for many models, not only in telecommunication.

In ▶ Chap. 5 we turn to a subject that is characteristic and indispensable for telecommunication applications, the *interference*. We introduce one of the most commonly used mathematical approaches: a success criterion for a one-hop message transmission in terms of the *signal-to-interference-and-noise ratio* (SINR). This introduces a much more realistic connectivity structure, but on the other hand creates a complex interaction in the point process. After we introduce this concept, we demonstrate some model calculations and give some results about the percolation behavior of the resulting random graph.

In ▶ Chap. 6 we develop the theory for the asymptotic analysis of two important asymptotic regimes: *high-density limits* (many devices in a bounded domain) and *spatial ergodic theorems* (many devices in a large domain with constant density). After a general discussion of the convergence of point processes, we give a useful basic result for the high-density setting. Then, we develop the *ergodic theory* of point process in large boxes. We begin with a reminder on the classical one-dimensional and discrete version, evolving around *Birkhoff's ergodic theorem*. Afterwards we formulate and prove its continuum, d-dimensional version, *Wiener's ergodic theorem*. In order to utilize this result for telecommunication questions, we invest some care into controlling boundary effects. Next, we introduce central notions of the ergodic theory like *ergodicity, mixing* and *metric transitivity*. Extensions to *marked* point processes and their Palm versions as well as to *empirical point fields* follow. Finally, we show how to apply the above theory to questions in the telecommunication context.

Furthermore, in ▶ Chap. 7 we introduce *large deviations theory*, which provides the basis for analyzing events with an extremely small probability as some parameter diverges. The exponential decay rate of the probability is described in terms of a variational formula that contains deeper information on the event and is amenable for further analysis. This theory does not belong to the core subjects of probability theory, its application to telecommunication systems is even less common, albeit highly relevant. After presenting an introductory example, we develop the general theory to the extent needed for subsequent applications. We concentrate on the two above-mentioned limiting situations and first analyze high-density limits for Poisson point processes in a fixed box. Based on this, we discuss in detail large deviations of three events, namely, the event of too many neighbors within a given distance, the event of bad connectivity in a single cell where messages are transmitted to the central base station with at most two hops, and the event of a large interference. Second, we turn our attention to the ergodic setting of large boxes. We present the general principle for the empirical stationary field of a marked Poisson point process and apply it to the event of atypically many disconnected transmitters.

In the final ▶ Chap. 8, for the first time, we go beyond the static setting and introduce time evolutions via propagation mechanisms for the messages in the multi-hop adhoc system. We provide a crash course on the classical theory of *interacting Markov jump processes* in continuous time, or *interacting particle systems* (IPS), which we extend towards random geometries. This enables us to study the spread of malware in device-to-device networks, more precisely, we present results about the propagation speed in the *Richardson model* based on the Boolean model. Next, we introduce two types of countermeasures that are intended to stop or at least slow down the malware spread. The first one is a rebooting mechanism that is mathematically equivalent to the *contact process* on the Boolean model for which we present a phase-transition result between survival and extinction. The second one consists of special devices, the *white knights*, which can eliminate the malware on neighboring devices. We exhibit regimes in which the addition of white knights leads to the elimination of the malware, both for fixed geometries and on the Boolean model. The functionality of the white knights as well as contact processes on random geometries are very recent additions to the literature of interacting particle systems and are presented here for the first time in textbook form.

The intention for writing this book was to give a concise and penetrable/comprehensible treatment, accessible for graduate students of mathematics with special focus on probability. We tried our best to illuminate the theory by providing a large number of use cases and we attempted to engage the reader inviting them to work on the many exercises that we collected. We believe that this book can serve as a solid basis for a 4 h lecture or a students' seminar in applied probability. Furthermore, it offers many opportunities for taking up motivating bachelor and master theses.

There are alternative introductions to the mathematical and in particular probabilistic treatment of spatial telecommunication systems. The most important ones are [BacBla09a, BacBla09b] and [FraMee08, Hän12], which all address different readerships and feature different aspects of the theory. Each of them served to some extent as sources for this book. In contrast to these textbooks, our text addresses mathematicians with a focus on probability theory, and we do not spend so much time on explicit calculations, but much more on concepts for understanding asymptotics and Markov process analysis. We are hardly touching topics such as transmission protocols, navigations and message routing strategies. On the other hand, as unique selling points, we offer material on random environments, rare events and random spread of malware.

Our own fascination for the subject started about 5 years ago when we began to discover that highly sophisticated probability theory can be utilized to help answer serious real-life questions in spatial telecommunication systems. We would feel blessed if this text helped to inspire probabilistically educated researchers to join us by employing their knowledge for analyzing such systems. We firmly believe that this field of activity has a bright future and will produce many deep and beautiful mathematical results for relevant questions. The joint venture between spatial probability and communication-network research has really just begun and offers a great variety of challenges for the future. Huge topics that we have not even touched are the consideration of message trajectories as stochastic processes on the network, random interference control for throughput optimization, mobility of devices and space-time evolutions of the entire system.

Device Locations: Point Processes

In this chapter, we introduce the basic mathematical model for the random locations of many point-like objects in the Euclidean space, the *Poisson point process (PPP)*. This process will be used for modeling the places of devices, additional boxes (supporting devices) and/or base stations in space. Apart from this interpretation in telecommunication, the PPP is universally applicable in many situations and is fundamental for the theory of stochastic geometry. The main assumption is a high degree of statistical independence of all the random points, which leads to many explicit and tractable formulas and to the validity of many properties that make a mathematical treatment simple. For these reasons, the PPP is the initial method of choice practically in any spatial telecommunication modeling and the most obvious starting point for a mathematical analysis.

PPPs belong to the core subjects of probability theory, and there is a number of general mathematical introductory and deepening texts on this subject, e.g., [Kin95] or [LasPen17], as well as texts with emphasis on applications in telecommunication, or chapters on PPPs, like [FraMee08, Hän12, Pen03]. Much more technical and comprehensive texts about general point processes are [DalVer03, DalVer08] and [Res87].

In ▶ Sect. 2.1, some of the groundwork will be presented within the general framework of point processes for which we set up only a minimal amount of measure-theoretic and topological structure. In ▶ Sect. 2.2, we introduce the PPP and many of its properties. The Cambell formula, an indispensable tool for handling the PPP, is presented in ▶ Sect. 2.3. An important enrichment of the modeling is introduced in ▶ Sect. 2.4, namely the marking of each of the points of the process with some individual random object. ▶ Section 2.5 explains how to give a mathematically rigorous meaning to taking the perspective from a given one of the points of the cloud, via Palm theory.

2.1 Point Processes

In this section, we introduce random point clouds as random variables and discuss briefly some basics on topology and measurability. See for example [DalVer03, Appendix A2] and [Res87] for details and proofs.

© Springer Nature Switzerland AG 2020
B. Jahnel, W. König, *Probabilistic Methods in Telecommunications*, Compact Textbooks in Mathematics,
https://doi.org/10.1007/978-3-030-36090-0_2

To begin with, we fix a dimension $d \in \mathbb{N}$ and a measurable set $D \subset \mathbb{R}^d$, which in our interpretation is the *communication area* where measurability on $\mathbb{R}^d$ is considered with respect to the Borel σ-algebra $\mathcal{B}(\mathbb{R}^d)$. In D, we assume that a point cloud $\phi = (x_i)_{i \in I}$, with an index set I, is given. This is interpreted as the cloud of the *locations of the devices* (e.g., devices, supporting devices, base stations, etc.). We would like to have that these locations do not coincide or accumulate anywhere in D, i.e., that $x_i \neq x_j$ for any $i \neq j$, and that any compact subset of D receives only finitely many of the x_i. Hence, the index set I is at most countable. Actually, we do not want to distinguish the points, but indeed consider only the *set* $\phi = \{x_i : i \in I\}$ or, equivalently, the *point measure* $\sum_{i \in I} \delta_{x_i}$. In other words, we would like to consider point clouds in the set

$$\mathbb{S}(D) = \{\phi \subset D : \#(\phi \cap A) < \infty \text{ for any bounded } A \subset D\}. \tag{2.1.1}$$

The elements of $\mathbb{S}(D)$ are called *locally finite sets*. We also call such an element a *point cloud* or a *point process* in D, and we will choose the perspective of ϕ being a point set $\{x_i : i \in I\}$, sometimes also written $(x_i)_{i \in I}$, or a point measure $\sum_{i \in I} \delta_{x_i}$ whenever convenient. Point measures ϕ of this form are called *simple* in the literature, since they satisfy $\phi(\{x\}) \in \{0, 1\}$ for any $x \in D$, by our assumption that the x_i are pairwise distinct. We will for the most part drop the term 'simple', as we will consider only simple ones.

Ultimately, we want to describe the distribution of a *random point cloud* $\Phi = (X_i)_{i \in I}$ in D with random index set I, which we call a *random point process*, i.e., a random variable with values in $\mathbb{S}(D)$. For this, we need a measurable structure on the state space $\mathbb{S}(D)$. We will now introduce a natural one, which also has the advantage that it comes from a topology as its Borel σ-algebra. Hence, it will be enough to introduce the topology. In order to do this, we assume the perspective of $\mathbb{S}(D)$ being the set of simple point measure and describe the topology by testing elements of $\mathbb{S}(D)$ against a suitable class of functions. More precisely, we consider functionals of the form

$$\begin{aligned} \phi(f) &= \int f(x) \, \phi(\mathrm{d}x) = \langle f, \phi \rangle \\ &= \int f(x) \sum_{i \in I} \delta_{x_i}(\mathrm{d}x) = \left\langle f, \sum_{i \in I} \delta_{x_i} \right\rangle = \sum_{i \in I} f(x_i), \end{aligned} \tag{2.1.2}$$

with $f : D \to \mathbb{R}$ taken from a suitable class of functions. In (2.1.2) we display a number of different notations for the integral of f with respect to ϕ that we will use throughout the book. The approach of taking integrals with respect to some family of test functions is an adaptation of the well-known characterization of the weak topology on the set of (probability) measures to the current setting of point measures. Note that point measures are in general not normalized and in fact often have infinite total mass, which would render $\phi(f)$ equal to ∞ for many functions f. It makes more sense to test the point cloud only in local areas, and this is what we want to do now. The set of test functions we use is

$$\mathcal{C}_{\mathrm{c}}(D) = \{f : D \to \mathbb{R} \mid f \text{ is continuous with compact support}\}.$$

Note that, in the definition of $\mathcal{C}_c(D)$, instead of a compact support, we could equivalently talk of a bounded support.

Definition 2.1.1 (Vague Topology on $\mathbb{S}(D)$)

The *vague topology* is the smallest topology on $\mathbb{S}(D)$ such that, for any function $f \in \mathcal{C}_c(D)$, the map $\phi \mapsto \phi(f)$ is continuous.

For a given point cloud $\phi = (x_i)_{i \in I} \in \mathbb{S}(D)$, we denote the number of its points in a given measurable set $A \subset D$ by

$$\phi(A) = \#\{i \in I : x_i \in A\} = \phi(\mathbb{1}_A) \in \mathbb{N}_0 \cup \{\infty\}, \tag{2.1.3}$$

where we write $\mathbb{1}_A(x) = 1$ if $x \in A$ and $\mathbb{1}_A(x) = 0$ otherwise for the indicator function on A. We want to properly speak about random point clouds and thus have to fix a σ-algebra on $\mathbb{S}(D)$.

Lemma 2.1.2 (σ-Algebra on $\mathbb{S}(D)$) *The Borel σ-algebra on $\mathbb{S}(D)$, based on the vague topology, is characterized as the σ-algebra generated by sets of the form $\{\phi \in \mathbb{S}(D) : \phi(A) = k\}$ for all measurable $A \subset D$ and $k \in \mathbb{N}_0$.*

In other words, the Borel σ-algebra on $\mathbb{S}(D)$, based on the vague topology, is the smallest σ-algebra on $\mathbb{S}(D)$, such that the maps $\phi \mapsto \phi(A)$ are measurable for all measurable $A \subset D$. We will in the following, if nothing else is stated, consider the Borel measurability induced by the vague topology and call this just *measurability*.

Remark 2.1.3 (σ-Algebras and Topologies)

1. *The τ-topology.* There are other useful topologies on $\mathbb{S}(D)$, for example the τ-*topology*, which is the smallest topology on $\mathbb{S}(D)$ such that, for any function $f \in \mathcal{T}(D)$, the map $\phi \mapsto \phi(f)$ is continuous where

$$\mathcal{T}(D) = \{f : D \to \mathbb{R} \mid f \text{ is measurable and bounded with compact support}\}.$$

   Obviously, every vaguely open set is also τ-open, i.e., the τ-topology is finer than the vague one. This implies that every $\mathbb{S}(D)$-valued random variable with respect to the Borel σ-algebra associated with the τ-topology is a random variable with respect to the Borel σ-algebra associated with the vague topology.

2. *Extension to topological spaces D.* One can easily extend the above topologies to locally compact topological spaces D, and we will make use of that later in ▶ Sect. 2.4.

3. *Extension to measurable spaces D.* Let us note that the theory of point processes can be developed in much greater generality (see, e.g., [LasPen17]), where instead of $(\mathbb{R}^d, \mathcal{B}(\mathbb{R}^d))$ a general measurable space $(W, \mathcal{W})$ is considered without any reference to topologies. Then, $\mathbb{S}(D)$ is usually replaced by the space **N** of all measures that can be written as a countable sum of measures ν with the property that $\nu(B) \in \mathbb{N}_0$ for all

$B \in \mathcal{W}$. A σ-algebra on $\mathbf{N}$ can then, for example, be defined via generating sets of the form $\{v \in \mathbf{N} : v(B) = k\}$ with $B \in \mathcal{W}, k \in \mathbb{N}_0$.   ◊

Exercise 2.1.4
Show that the map $D \to \mathbb{S}(D), x \mapsto \delta_x$, is measurable.   ◊

From the characterization of the σ-algebra in Lemma 2.1.2, we can derive the following characterization of the distribution of an $\mathbb{S}(D)$-valued random variable.

Lemma 2.1.5 (Characterization of Distributions) *The distribution of an $\mathbb{S}(D)$-valued random variable Φ is uniquely determined by the distributions of all vectors $(\Phi(A_1), \ldots, \Phi(A_k))$ with $k \in \mathbb{N}$ and measurable bounded sets $A_1, \ldots, A_k \subset D$.*

In analogy with stochastic processes with parameter set $\mathbb{N}_0$ instead of D, one can see these vectors as defining the finite-dimensional distributions of the point process Φ.

Another important characterization of the distribution of an $\mathbb{S}(D)$-valued random variable $\Phi = (X_i)_{i \in I}$ is in terms of its *Laplace transform* defined by

$$\mathcal{L}_\Phi(f) = \mathbb{E}\left[\exp\left(-\sum_{i \in I} f(X_i) \right) \right] \in [0, 1], \quad f : D \to [0, \infty) \text{ measurable.} \quad (2.1.4)$$

Note that here $\mathcal{L}_\Phi$ does not depend on Φ, but on the distribution of Φ.

Lemma 2.1.6 (Laplace Transform Fixes Distributions) *The distribution of an $\mathbb{S}(D)$-valued random variable Φ is uniquely determined by its Laplace transform for all measurable nonnegative functions f with compact support.*

For describing the distribution of a random point cloud, it appears natural to do this in terms of a measure μ on D, which gives a first rough idea how many points of Φ are located in a given set.

Definition 2.1.7 (Intensity Measure)

The *intensity measure* μ of a random point cloud Φ in D is defined by

$$\mu(A) = \mathbb{E}[\Phi(A)] \in [0, \infty], \quad A \subset D \text{ measurable.} \quad (2.1.5)$$

However, the intensity measure is by far not enough to fully characterize the distribution of the random point process Φ. In particular, it does not reveal anything about the spatial correlation structure of Φ. Similar to the characterization of the distribution of random variables via its moments, let us mention that point processes can be characterized by their *factorial moment measures*, for details see for example [DalVer03].

An important class of random point clouds are the stationary ones.

Definition 2.1.8 (Stationary Random Point Clouds) ─────────────

A random point cloud Φ in $\mathbb{R}^d$ (more precisely, its distribution) is called *stationary* if its distribution is identical to the one of $\Phi + x$ for any $x \in \mathbb{R}^d$.

Exercise 2.1.9

Show that for all $x \in \mathbb{R}^d$, the shift $\Phi \mapsto \Phi + x$ as well as the addition of the point $\Phi \mapsto \Phi + \delta_x$ are measurable. ◇

Observe that the term 'stationarity' makes sense only for random point processes on $D = \mathbb{R}^d$. Sometimes, the notion *homogeneous* is used instead of a stationary one. The intensity measure of a stationary point cloud is invariant under shifts and therefore equal to a multiple of the Lebesgue measure dx on $\mathbb{R}^d$:

Lemma 2.1.10 (Intensity of Stationary Point Processes) *Let Φ be a stationary point process on $\mathbb{R}^d$ with intensity measure μ such that $\mu([0, 1]^d) < \infty$. Then $\mu(dx) = \lambda dx$ with $\lambda = \mu([0, 1]^d)$.*

If Φ is a stationary PPP with intensity measure λdx, then we call λ its *intensity*.

Exercise 2.1.11

Show that a stationary point process Φ on $\mathbb{R}^d$ cannot have a positive but finite total number of points with positive probability. *Hint:* Assume $\mathbb{P}(0 < \Phi(\mathbb{R}^d) < \infty) > 0$, condition on this event and find a contradiction to stationarity. ◇

Finally let us mention another notion about random point processes capturing possible fixed locations.

Definition 2.1.12 (Fixed Atoms) ──────────────

A random point process Φ on D is said to have a *fixed atom* at $x \in D$ if $\mathbb{P}(\Phi(\{x\}) > 0) > 0$. A random point process without fixed atoms is called *atomless*.

Fixing atoms can be useful, for example, if devices can only appear on a discrete subset of $\mathbb{R}^d$.

Exercise 2.1.13

Show that for any stationary point process with positive intensity, the probability of the event that the origin has two or more distinct points with the same distance to the origin, is zero. ◇

2.2   Poisson Point Processes

In this section, we introduce a very particular random point process that is characterized by a very high degree of independence.

Definition 2.2.1 (Poisson Point Process)

Let μ be a measure on D. We call the random point process Φ a *Poisson point process (PPP)* with *intensity measure* μ if, for any $k \in \mathbb{N}$ and any pairwise disjoint measurable sets $A_1, \ldots, A_k \subset D$ with finite μ-measure, the counting variables $\Phi(A_1), \ldots, \Phi(A_k)$ are independent Poisson-distributed random variables with parameters $\mu(A_1), \ldots, \mu(A_k)$, i.e., if for all $n_1, \ldots, n_k \in \mathbb{N}_0$,

$$\mathbb{P}\big(\Phi(A_1) = n_1, \ldots, \Phi(A_k) = n_k\big) = \prod_{i=1}^{k} \left[e^{-\mu(A_i)} \frac{\mu(A_i)^{n_i}}{n_i!} \right]. \tag{2.2.1}$$

In ◻ Fig. 2.1 we present a realization of a PPP. Let us further make a number of comments.

Remark 2.2.2   (First Properties of PPPs)

1. *Definition on $\mathbb{R}^d$.* We can certainly also drop the set D, i.e., put it equal to $\mathbb{R}^d$, since the dependence on D can be absorbed into μ. Indeed, a measure μ on D can be trivially extended to a measure on $\mathbb{R}^d$ by assigning the value zero on $D^c = \mathbb{R}^d \setminus D$. Then the PPP that is induced by μ on D and the one that is induced by its extension on $\mathbb{R}^d$ are equal to each other in distribution, after restricting to D.

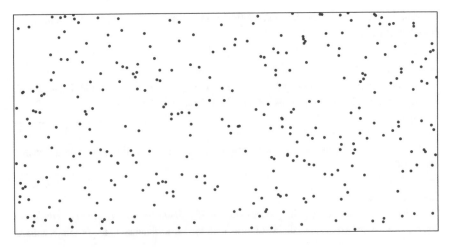

◻ **Fig. 2.1**   Realization of a stationary PPP

2. *Intensity measure unique for PPP.* From Lemma 2.1.5 it follows that each measure on $\mathbb{R}^d$ is the intensity measure of a unique PPP, up to distribution.

3. *Arbitrary measurable spaces.* It is clear that the space D (or $\mathbb{R}^d$) can be widely generalized for Definition 2.2.1 to make sense; in fact we need only a measure μ on an arbitrary measurable space D that can be written as a countable sum of finite measures on D, see, e.g., [LasPen17]. For our setting of point processes, based on (2.1.1), we need to assume that D is a locally compact topological space.

4. *Intensity measures.* What we called the intensity measure of a PPP in Definition 2.2.1 is consistent with Definition 2.1.7, as is seen easily. ◇

At this point, let us recall and collect some important properties of the Poisson distribution.

Remark 2.2.3 (Poisson Distributions) The *Poisson distribution* with parameter $\alpha \in (0, \infty)$, or a Poisson-distributed random variable N with parameter α, is given by

$$\mathrm{Po}_\alpha(k) = \mathbb{P}(N = k) = \mathrm{e}^{-\alpha} \frac{\alpha^k}{k!}, \qquad k \in \mathbb{N}_0.$$

It has expectation and variance equal to α. Here are more interesting properties.

1. *Generating function.* Its generating function is given by $\sum_{k \in \mathbb{N}_0} \mathrm{Po}_\alpha(k) s^k = \mathbb{E}[s^N] = \exp(-(1-s)\alpha)$ for $s \in [0, \infty)$.

2. *Laplace transform.* For $r \in [0, \infty)$, its Laplace transform is given by $\mathcal{L}_\alpha(r) = \sum_{k \in \mathbb{N}_0} \mathrm{Po}_\alpha(k) \exp(-rk) = \mathbb{E}[\exp(-rN)] = \exp\{\alpha(\exp(-r) - 1)\}$.

3. *Sums of Poisson random variables.* The sum of an arbitrary finite number of independent Poisson-distributed random variables is also Poisson-distributed, and the parameter is the sum of the parameters of the summands.

4. *Poisson limit theorem.* The distribution of a sum of n independent binomially distributed random variables with parameter α/n converges towards Po_α as $n \to \infty$.

5. *Thinning of Poisson distributions.* Given a Poisson-distributed random variable N with parameter α and independent and identically distributed (i.i.d.) Bernoulli random variables $Y_1, \ldots, Y_N$ with parameter $p \in [0, 1]$, then $\sum_{i=1}^N Y_i$ is Poisson-distributed with parameter $p\alpha$.

6. *Sums of exponential random variables.* If N is Poisson distributed with parameter αt where $\alpha, t > 0$, then $\mathbb{P}(N \geq k) = \mathbb{P}(E_1 + \cdots + E_k \leq t)$ for any $k \in \mathbb{N}$, where $(E_i)_{i \in \mathbb{N}}$ is a sequence of i.i.d. random variables having the exponential distribution with parameter α.

The *superposition principle (Sums of Poisson random variables, the third property above).* is the main reason why the Poisson distribution is the 'right' distribution for the PPP, which can be seen by considering the distribution of the PPP on non-disjoint sets. ◇

Exercise 2.2.4

Prove the statements of Remark 2.2.3. ◇

Exercise 2.2.5

Let $F: \mathbb{N}_0 \to \mathbb{R}$ be bounded and N_α a Poisson-distributed random variable with parameter $\alpha > 0$. Calculate the derivative $\frac{d}{d\alpha}\mathbb{E}[F(N_\alpha)]$.   ◊

Definition 2.2.1 is formulated in terms of the distribution of the counting variables $\Phi(A)$, but we would like to have the object Φ also as an explicit $\mathbb{S}(D)$-valued random variable. This is provided by the following construction.

Lemma 2.2.6 (Construction of PPPs) *Assume that μ is a measure on D with $\mu(D) \in (0, \infty)$. Let $N(D)$ be a Poisson random variable with parameter $\mu(D)$. Put $I = \{1, \ldots, N(D)\}$. Given $N(D)$, let $\Phi = (X_i)_{i \in I}$ be a collection of independent random points in D with distribution $\mu(\cdot)/\mu(D)$. Then Φ is a PPP with intensity measure μ.*

This construction works a priori only for finite intensity measures, but if μ is infinite and σ-finite, then one can decompose D into countably many measurable sets with finite μ-measure, construct the random point process on the partial sets according to Lemma 2.2.6 independently and put all these point processes together in order to obtain a PPP with intensity measure μ on D. It is an exercise to show that this construction works and that the resulting random point process is independent of the chosen decomposition of D; this is basically the proof of Lemma 2.2.13 below.

Exercise 2.2.7

Show that if μ has a Lebesgue density, then the corresponding PPP is simple.   ◊

Exercise 2.2.8

Show that for a PPP Φ on D with intensity measure μ, to have an atom at $x \in D$, is equivalent to $\mu(\{x\}) > 0$.   ◊

Exercise 2.2.9

Let Φ be a PPP on D with intensity measure μ such that $\mu(D) \in (0, \infty)$. Show that for all measurable $B \subset D$

$$\mathbb{P}\big(\Phi(B) = k \mid \Phi(D) = n\big) = \binom{n}{k}p^k(1-p)^{n-k}, \qquad 0 \leq k \leq n, k, n \in \mathbb{N}_0,$$

where $p = \mu(B)/\mu(D)$. In other words, conditioned on the total number of Poisson points in D, the distribution of $\Phi(B)$ is binomial.   ◊

Exercise 2.2.10 (Further Characterizations of PPPs)

1. Show that a random point process Φ is a PPP if and only if there exists a σ-finite measure μ on $\mathbb{R}^d$ such that $\mathbb{P}(\Phi(A) = n) = e^{-\mu(A)}\mu(A)^n/n!$ for all bounded and measurable $A \subset \mathbb{R}^d$ and $n \in \mathbb{N}_0$.
2. Show that a random point process Φ is a PPP if and only if there exists a σ-finite measure μ on $\mathbb{R}^d$ such that $\mathbb{P}(\Phi(A) = 0) = \exp(-\mu(A))$ for all bounded and measurable $A \subset \mathbb{R}^d$.

3. Show that if Φ is atomless, then Φ is a PPP if and only if Φ is simple and completely independent, i.e., $\Phi(A_1), \ldots, \Phi(A_n)$ are independent for all $n \in \mathbb{N}$ and disjoint bounded and measurable $A_1, \ldots, A_n \subset \mathbb{R}^d$. $\Diamond$

Example 2.2.11 (PPPs)

1. *Standard PPPs.* The *standard PPP* on $D = \mathbb{R}^d$ is obtained for the intensity measure λdx, where $\lambda \in (0, \infty)$ is the *intensity*. Since dx is a stationary measure, the corresponding random point process is stationary (see Definition 2.1.8) and often referred to as a *homogeneous PPP*. Actually, since λdx is the only shift-invariant measure on $\mathbb{R}^d$, every stationary PPP has this as its intensity measure for some $\lambda \in (0, \infty)$, see Lemma 2.1.10. Furthermore, its distribution is also *isotropic*, i.e., rotationally invariant.

2. *Discrete PPPs.* For $D = \mathbb{Z}^d$ and μ the counting measure on D, the corresponding PPP is a discrete variant of the standard PPP. It is obtained by realizing independent and identically distributed Poisson variables for each $x \in \mathbb{Z}^d$ and putting that number of points into x. Alternatively, one can, for any finite set $\Lambda \subset \mathbb{Z}^d$, generate a Poisson random variable $N(\Lambda)$ with parameter $\#\Lambda$, and distribute $N(\Lambda)$ points independently and uniformly over Λ, decompose $\mathbb{Z}^d$ into such sets and add all these points independently in all of $\mathbb{Z}^d$; the resulting superposition is the desired PPP. Note that the PPP on $\mathbb{Z}^d$ is not simple and has fixed atoms in each $x \in \mathbb{Z}^d$. Actually, we will never be working with a discrete PPP in the following.

3. *Modeling using PPPs.* If we want to model the locations of devices in a given city D, then the Lebesgue density of μ should be low in low density areas like lakes, forests and fields and high in high density areas like highly frequented places.

4. *PPPs in time.* The one-dimensional standard PPP has enormous importance in the modeling of random times, in particular in the theory of time-homogeneous Markov chains in continuous time. For $d = 1$, another interesting characterization of the PPP is possible: the distances of neighboring pairs of points of the PPP are i.i.d. exponentially distributed random variables with the same parameter as the PPP (see also the last property that we mention in Remark 2.2.3). This is directly connected with the famous property of *memorylessness* of the process of times at which the points appear, see also ▶ Chap. 8. However, we will not elaborate on these nice properties here, since we are mainly interested in $d \geq 2$. $\Diamond$

Example 2.2.12 (Contact Distance)

The *contact distance* of a space point $x \in D$ to a set $\phi = \{x_i : i \in I\} \in \mathbb{S}(D)$ is defined by

$$\text{dist}(x, \phi) = \inf\{|x - x_i| : i \in I\}. \tag{2.2.2}$$

This is the radius of the largest open ball around x that contains no point of ϕ. If Φ is a PPP with intensity measure μ, then the distribution function of $\text{dist}(x, \Phi)$ is easy to find:

$$\mathbb{P}\big(\text{dist}(x, \Phi) < r\big) = 1 - \mathbb{P}\big(\Phi(B_r(x)) = 0\big) = 1 - \exp\big(-\mu(B_r(x))\big), \tag{2.2.3}$$

where $B_r(x)$ is the open ball with radius r, centered at $x \in \mathbb{R}^d$. $\Diamond$

Let us derive some important and simple properties of PPPs. First, we identify the distribution of the superposition of independent processes.

Lemma 2.2.13 (Superposition of PPPs) *Let μ and ν be two measures on D and let $\Phi = (X_i)_{i \in I}$ and $\Psi = (Y_i)_{i \in J}$ be independent PPPs with intensity measures μ and ν, respectively. Then, $\{X_i : i \in I\} \cup \{Y_i : i \in J\}$ is a PPP with intensity measure $\mu + \nu$.*

The proof uses the well-known property of the sum of independent Poisson random variables to be again Poisson, see Remark 2.2.3. An extension to superpositions of countably many independent PPPs is straightforward.

Exercise 2.2.14
Prove Lemma 2.2.13 and its extension to countable unions.   ◊

Another operation that goes well with PPPs is random thinning, i.e., the random removal of some of the points.

Lemma 2.2.15 (Random Thinning of PPPs) *Let $\Phi = (X_i)_{i \in I}$ be a PPP in D with intensity measure μ. With a probability $p \in [0, 1]$, given Φ, we keep independently any of the particles X_i. Then, the remains are a PPP with intensity measure $p\mu$.*

Also proving this is an elementary exercise, which is based on Property 5 in Remark 2.2.3.

In the context of telecommunications, for example in the *ALOHA protocol*, each device independently chooses to take part in the network at a given time instant (see ▶ Sect. 5.4 for some more comments). If these choices are identically distributed and the device positions follow a PPP, then according to Lemma 2.2.15, the process of active devices is again a PPP.

Exercise 2.2.16
Prove Lemma 2.2.15 and an extension in which p is a measurable function $x \mapsto p(x) \in [0, 1]$ and the particle X_i is removed with probability $p(X_i)$. Show that the thinned PPP has intensity $p(x)\mu(\mathrm{d}x)$.   ◊

Exercise 2.2.17
Let Φ and Ψ be two independent PPPs with intensity measures μ and ν. Show that Φ has the same distribution as the p-thinning of the joint process $\Phi + \Psi$ where $p(x) = \mathrm{d}\mu/\mathrm{d}(\mu + \nu)(x)$.   ◊

Here is another, quite general, way to construct a PPP out of another one, by a measurable mapping.

Theorem 2.2.18 (Mapping Theorem)
Let $\Phi = (X_i)_{i \in I}$ be a PPP with intensity measure μ on $D \subset \mathbb{R}^d$, and let $f : D \to \mathbb{R}^s$ be a measurable map. Then, $f(\Phi) = (f(X_i))_{i \in I}$ is a PPP in $\mathbb{R}^s$ with intensity measure $\mu \circ f^{-1}$.

Let us note that, in the context of telecommunications, in order to avoid that the image PPP has fixed atoms, we must assume that $\mu(f^{-1}(\{x\})) = 0$ for any $x \in \mathbb{R}^s$. The proof is an easy application of Campbell's theorem, which we present in the next section.

Exercise 2.2.19
Let Φ be an inhomogeneous PPP on $[0, \infty)$ with intensity measure μ. Find a mapping $f : [0, \infty) \to [0, \infty)$ such that $f(\Phi)$ is a homogeneous PPP on $[0, \infty)$. ◇

The following formula provides an explicit expression for the construction of a PPP as presented in Lemma 2.2.6.

Lemma 2.2.20 (Explicit Representation) *Let Φ be a PPP on D with intensity measure μ such that $\mu(D) \in (0, \infty)$. Then, for any measurable function $f : \mathbb{S}(D) \to [0, \infty)$,*

$$\mathbb{E}[f(\Phi)] = e^{-\mu(D)} f(\emptyset) + e^{-\mu(D)} \sum_{n \in \mathbb{N}} \frac{1}{n!} \int_{D^n} f(\{x_1, \dots, x_n\}) \, \mu^{\otimes n}(\mathrm{d}(x_1, \dots, x_n)).$$

In some situations, the following fact about the covariance of counting variables of points in a PPP is useful.

Lemma 2.2.21 (Covariances) *Let Φ be a PPP on D with intensity measure μ, and let A_1 and A_2 be two measurable subsets of D with $\mu(A_1), \mu(A_2) < \infty$, then the covariance of $\Phi(A_1)$ and $\Phi(A_2)$ is equal to $\mu(A_1 \cap A_2)$.*

Exercise 2.2.22
Prove Lemma 2.2.20 and Lemma 2.2.21. ◇

2.3 Campbell Moment Formulas

As always, we let $D \subset \mathbb{R}^d$ be a measurable (bounded or unbounded) subset of $\mathbb{R}^d$, the communication area. In Sect. 2.1, we defined a topology on $\mathbb{S}(D)$ by means of the maps $\phi \mapsto \phi(f) = \sum_{i \in I} f(x_i)$ (see (2.1.2)) for certain functions $f : D \to \mathbb{R}$. Hence, the expectation $\mathbb{E}[\Phi(f)]$ will be an important tool for characterizing the distribution of a random point process Φ. Below, we will give a formula for this expression.

Furthermore, the Laplace transform $\mathcal{L}_\Phi(f)$ (see (2.1.4)) turned out in Lemma 2.1.6 to uniquely determine the distribution of a random point process Φ, hence it will also be useful to have explicit formulas for $\mathcal{L}_\Phi(f)$. This functional has the great advantage that it always yields a finite value for nonnegative functions f and has very nice properties with respect to convergence of the random point process, as an application of the bounded convergence theorem is always possible, see ▶ Chap. 6. The next theorem will also give a handy formula for this in the case of a PPP.

Theorem 2.3.1 (Campbell's Theorem)

Let Φ be a random point process on D with intensity measure μ, and let $f : D \to \mathbb{R}$ be integrable with respect to μ or non-negative and measurable, then

$$\mathbb{E}[\Phi(f)] = \int_D f(x)\,\mu(\mathrm{d}x). \tag{2.3.1}$$

If Φ is even a PPP, then for nonnegative f,

$$\mathcal{L}_\Phi(f) = \mathbb{E}[e^{-\Phi(f)}] = \exp\left(\int_D (e^{-f(x)} - 1)\,\mu(\mathrm{d}x)\right). \tag{2.3.2}$$

Proof
The first statement (2.3.1) can be essentially seen as a consequence of Fubini's theorem. More precisely, for $f = \alpha \mathbb{1}_A$ with measurable $A \subset D$ and $\alpha \in \mathbb{R}$, we have

$$\mathbb{E}[\Phi(f)] = \alpha\mathbb{E}[X(A)] = \alpha\mu(A) = \int_D f(x)\,\mu(\mathrm{d}x)$$

and

$$\mathcal{L}_\Phi(f) = \mathbb{E}[\exp(-\alpha\Phi(A))] = \exp\left((\exp(-\alpha) - 1)\mu(A)\right),$$

where we use the formula for the Laplace transform of the Poisson random variable $\Phi(A)$, see Remark 2.2.3. Now by standard measure-theoretic arguments, using monotonicity and linearity, these statements can be extended to all integrable non-negative and measurable functions f. □

Exercise 2.3.2
Let Φ be a homogeneous PPP. Show that $\Phi(f) = \sum_{i \in I} f(X_i)$ is absolutely convergent almost surely if and only if $\int \min(|f(x)|, 1)\,\mathrm{d}x < \infty$. ◊

Exercise 2.3.3
Let Φ be a homogeneous PPP with intensity $\lambda > 0$. Prove that

$$\frac{\mathrm{d}}{\mathrm{d}\lambda}\mathbb{E}[\Phi(f)] = \int f(x)\,\mathrm{d}x$$

for all Lebesgue integrable $f : \mathbb{R}^d \to \mathbb{R}$. ◊

As an application, let us present here the proof of the Mapping Theorem 2.2.18 from the previous section.

Proof of the Mapping Theorem 2.2.18

For all non-negative measurable functions g, the Laplace transform of $f(\Phi)$ at g is given by

$$\mathcal{L}_{f(\Phi)}(g) = \mathbb{E}\Big[\exp\Big(-\sum_{i \in I} g(f(X_i))\Big)\Big] = \exp\Big(\int \big(\exp(g(f(x))) - 1\big)\,\mu(\mathrm{d}x)\Big),$$

see Campbell's theorem. But this is the Laplace transform at g of a PPP with intensity measure $\mu \circ f^{-1}$. Since the Laplace transform fixes distributions, see Lemma 2.1.6, the proof is finished. □

2.4 Marked Poisson Point Processes

To each of the points of a point process, we may add some individual information, which may also be random. In a model for the locations of the devices, this may be the strength of the transmitted signal sent out from the location of this device, or the fact of whether the device is sending or receiving, and much more. For example, in ▶ Chap. 4 we will attach to each device its coverage zone given by a disc with some fixed or random radius, constituting the famous Boolean model.

We call this additional information the *mark* of the device. Hence, a *marked point process* in D is essentially nothing but a random element of $\mathbb{S}(D \times \mathcal{M})$, where $\mathcal{M}$ is the set of marks, and we extended the definition of $\mathbb{S}(D)$ in (2.1.1) in an obvious way. We will write an element of $\mathbb{S}(D \times \mathcal{M})$ always as set $\{(x_i, m_i) : i \in I\}$, as collection $(x_i, m_i)_{i \in I}$ or as the point measure $\sum_{i \in I} \delta_{(x_i, m_i)}$ and call m_i the mark of x_i. However, note that D and $\mathcal{M}$ play slightly different roles, as we require that the points x_i are pairwise different, but the marks do not have to be. In order to be able to use the notion of (2.1.1) in this way, we need to give $\mathcal{M}$ a topological structure and equip it with the corresponding Borel σ-algebra $\mathcal{B}(\mathcal{M})$. In order that we can use the topology introduced in ▶ Sect. 2.1, we also assume that $\mathcal{M}$ is locally compact, see Part 4 in Remark 2.1.3. We call $(\mathcal{M}, \mathcal{B}(\mathcal{M}))$ the *mark space*.

Now we introduce *random* marked point processes. These are essentially just random variables with values in the set $\mathbb{S}(D \times \mathcal{M})$. In order to avoid some technicalities, we restrict in this section to a particular class (which nevertheless gives rise to many interesting examples), the marked PPPs with independent marks. (In ▶ Sect. 6.6 we will extend the scope to a larger class in $D = \mathbb{R}^d$ and will discuss more examples.) Indeed, given a PPP $(X_i)_{i \in I}$ with intensity measure μ, we will attach to each X_i independently a mark M_i that may depend on the location of X_i. For doing this in a mathematically correct way, we need a probability kernel $K : D \times \mathcal{B}(\mathcal{M}) \to [0, 1]$, i.e., a map such that $K(x, \cdot)$ is a probability measure on $(\mathcal{M}, \mathcal{B}(\mathcal{M}))$ for any $x \in D$, and $K(\cdot, G)$ is measurable for any $G \in \mathcal{B}(\mathcal{M})$. Then $K(x_i, \cdot)$ is the distribution of the mark that is attached to the point x_i, and it may therefore depend on this point, but not on the index i. We obtain the following notion.

Definition 2.4.1 (Marked PPP)

Let $\Phi = (X_i)_{i \in I}$ be a PPP in D with finite intensity measure μ, and let $(\mathcal{M}, \mathcal{B}(\mathcal{M}))$ be a measurable space, the *mark space*. Furthermore, let K be a probability kernel from D to $\mathcal{M}$. Given Φ, let $(M_i)_{i \in I}$ be an independent collection of $\mathcal{M}$-valued random variables with distribution $\otimes_{i \in I} K(X_i, \cdot)$ (where the i-th factor acts on M_i). Then, the point process $\Phi_K = ((X_i, M_i))_{i \in I}$ in $D \times \mathcal{M}$ (the random point measure $\sum_{i \in I} \delta_{(X_i, M_i)}$) is called a *K-marked Poisson point process (K-MPPP)* or a *K-marking* of the PPP Φ.

We call $(M_i)_{i \in I}$ also an *independent marking* since the family of marks $(M_i)_{i \in I}$ is independent given Φ. If K does not depend on the first argument, then $(M_i)_{i \in I}$ is even an *i.i.d.* collection given Φ, and we refer to this marking as an *i.i.d. marking*. It is not a problem to also handle non-independent markings. As an example, we later consider a model where we attach to each device X_i a coverage zone given by its Voronoi cell $z(X_i) = z(X_i, \Phi)$, as defined in (3.2.1), which shows a non-local long-range dependence on the point configuration surrounding X_i. Many more examples for K-MPPPs in $\mathbb{R}^d$ with telecommunication applications are presented in ▶ Sects. 6.6 and 6.8.

Let us calculate the Laplace transform of a K-MPPP.

Lemma 2.4.2 (Laplace Transform of a K-MPPP) *For all measurable and compactly supported $g \colon D \times \mathcal{M} \to [0, \infty)$, the Laplace transform of the K-marking Φ_K in Definition 2.4.1 is given by*

$$\mathcal{L}_{\Phi_K}(g) = \mathcal{L}_\Phi(g^*), \tag{2.4.1}$$

where

$$g^*(x) = -\log\left(\int_{\mathcal{M}} \exp(-g(x, y)) \, K(x, \mathrm{d}y)\right), \qquad x \in D. \tag{2.4.2}$$

Proof
We use the so-called tower property of conditional expectations and the independence over i to see that

$$\mathcal{L}_{\Phi_K}(g) = \mathbb{E}\Big[e^{-\sum_{i \in I} g(X_i, M_i)} \Big] = \mathbb{E}\Big[\prod_{i \in I} e^{-g(X_i, M_i)} \Big] = \mathbb{E}\Big[\mathbb{E}\Big[\prod_{i \in I} e^{-g(X_i, M_i)} \Big| \Phi \Big] \Big]$$

$$= \mathbb{E}\Big[\prod_{i \in I} \mathbb{E}\big[e^{-g(X_i, M_i)} \mid X_i \big] \Big].$$

Observe that

$$\mathbb{E}\big[\exp(-g(X_i, M_i))\big|X_i\big] = \int_{\mathcal{M}} K(X_i, dy)\, \exp(-g(X_i, y)) = \exp(-g^*(X_i)).$$

Substituting this yields the assertion. □

Now let us come back to the assumption that the mark space $\mathcal{M}$ is a locally compact topological space, see the remarks at the beginning of this section. Then Lemma 2.4.2 easily implies that the K-MPPP Φ_K is nothing but a usual PPP on $D \times \mathcal{M}$ with intensity measure $\mu \otimes K$,[1] where we slightly extended Definition 2.2.1 in the spirit of Part 3 of Remark 2.2.2.

> **Theorem 2.4.3 (Marking Theorem)**
> *Consider the situation of Definition 2.4.1 and assume that $(\mathcal{M}, \mathcal{B}(\mathcal{M}))$ is locally compact and is equipped with the Borel σ-algebra. Then Φ_K is in distribution equal to the PPP on $D \times \mathcal{M}$ with intensity measure $\mu \otimes K$.*

Proof

Using an extension of Lemma 2.1.6, we know that the distribution of a point process in $D \times \mathcal{M}$ is uniquely determined by its Laplace transform. Hence, we only have to show that the Laplace transform of a PPP with intensity measure $\mu \otimes K$ is identical to the one of a K-MPPP. We apply (2.3.2) to Lemma 2.4.2 to see that

$$\mathcal{L}_\Phi(g^*) = \exp\left(\int_D (e^{-g^*(x)} - 1)\, \mu(dx)\right)$$

$$= \exp\left(\int_D \left(\int_{\mathcal{M}} K(x, dy)\, e^{-g(x,y)} - 1\right)\mu(dx)\right)$$

$$= \exp\left(\int_D \int_{\mathcal{M}} (e^{-g(x,y)} - 1)\, \mu(dx) K(x, dy)\right)$$

$$= \exp\left(\int_{D\times\mathcal{M}} (e^{-g} - 1)\mathrm{d}(\mu \otimes K)\right).$$

Now consulting (2.3.2) once more (for $D \times \mathcal{M}$ instead of D) we see that this is the Laplace transform of a PPP with intensity measure $\mu \otimes K$. □

Having seen this, it is also clear that it is not necessary to normalize K (as long as each $K(x, \cdot)$ is a finite measure, like μ), since one can construct a realization of such a K-MPPP also by first taking Φ as a PPP with intensity measure $\widetilde{\mu}(dx) =$

---

[1] Note that the measure $\mu \otimes K$ is defined by $\mu \otimes K(B) = \int_{B^{(1)}} \mu(dx) K(x, B_x^{(2)})$, where $B^{(1)} = \{x \in D: \exists y \in \mathcal{M}: (x, y) \in B\}$ and $B_x^{(2)} = \{y \in \mathcal{M}: (x, y) \in B\}$.

$K(x, \mathcal{M}) \, \mu(\mathrm{d}x)$ and then picking the marks with distribution $K(x, \cdot)/K(x, \mathcal{M})$. The reason is that, for any $c \in (0, \infty)$, the measures $c\mu(\mathrm{d}x) \otimes K(x, \mathrm{d}y)/c$ and $\mu(\mathrm{d}x) \otimes K(x, \mathrm{d}y)$ on $D \times \mathcal{M}$ coincide. It is also easy to see that, for any K-MPPP $((X_i, M_i))_{i \in I}$ with intensity measure μ and normalized mark kernel K, the projected process $(X_i)_{i \in I}$ is a PPP with intensity measure μ.

Exercise 2.4.4

Prove the last statement.   ◊

Let us present some applications of the Marking Theorem 2.4.3.

Remark 2.4.5 (Realization of Superpositions) With the help of the Marking Theorem 2.4.3, we can now easily realize many PPPs on one probability space with many different intensity measures, in other words, couple them. More precisely, let μ be a measure on D with $\mu(D) \in (0, \infty)$, then we can, for any $p \in [0, 1]$, construct the PPPs with intensity measure $p\mu$ on one probability space as follows. Consider the independently marked PPP $\Phi = ((X_i, U_i))_{i \in I}$ with intensity measure $\mu \otimes U$ where U denotes the uniform distribution on $[0, 1]$. Then $\{X_i : U_i \leq p\}$ is a PPP with intensity measure $p\mu$.

Remark 2.4.6 (Random Displacements and Dynamics of PPPs) The Marking Theorem 2.4.3 can be used to extend the Mapping Theorem 2.2.18 to the situation where the displacement function f is random, giving rise to a *displacement theorem*. More precisely, instead of f we consider a probability kernel $K : D \times \mathcal{B}(\mathbb{R}^s) \to [0, 1]$ where $K(x, \cdot)$ is the distribution of the random displacement of the point x. Consider the K-marking $((X_i, Y_i))_{i \in I}$ of the PPP $(X_i)_{i \in I}$ with intensity measure μ; i.e., Y_i is the displacement of X_i. We call $(Y_i)_{i \in I}$ the K-displaced process. Then it is a simple consequence of the marking theorem that it is again a PPP and has the intensity measure $\mu K(\cdot) = \int \mu(\mathrm{d}x) K(x, \cdot)$, see also [Kin95, Section 5.5].

This observation is highly interesting if we want to consider *Markovian dynamics* of PPPs. Indeed, if K is a kernel from D to D, then K can be seen as a random (discrete-time) Markovian movement scheme, which can be iterated: each point of the PPP independently performs this Markovian dynamics. The entire collection of these Markov chains in D is itself a Markov process in $\mathbb{S}(D)$, and it is at any time a PPP. Its intensity measure after the k-th iteration is $\mu K^{\otimes k}$, where $K^{\otimes k}$ is the k-th concatenation of K. Certainly, the K-marking can be applied to any point process in D, but the nice identification at any later time is due to the Poisson structure.   ◊

Exercise 2.4.7

Prove Remark 2.4.5 and Remark 2.4.6.   ◊

Example 2.4.8 (Dimensions as Marks)

Since the d-dimensional Lebesgue measure is the d-fold product measure of the one-dimensional Lebesgue measure, one could think that the standard PPP in $\mathbb{R}^d$ can be seen as a marked PPP in $\mathbb{R}^{d-1}$ with marks in $\mathbb{R}$. However, since the Lebesgue measure on $\mathbb{R}$ is not finite, this is not covered by Definition 2.4.1. If the last factor $\mathbb{R}$ is replaced by some bounded measurable set and the Lebesgue measure by the restriction, then this interpretation is correct.   ◊

Example 2.4.9 (Marks and the Matérn Hard-Core Point Processes)

To avoid, for example, interference between devices in a communication network, see ▶ Sect. 5.4, it can be desirable to model a point process with some hard-core repulsion. For example, one can impose the condition that any two points should have a minimal distance of $r > 0$. There are many different random point processes obeying this property with probability one. Maybe the most naive way of achieving this goal is to consider the image distribution of a PPP under the thinning map

$$\phi \mapsto \{x_i \in \phi \colon |x_i - x_j| \geq r \text{ for all } x_j \in \phi \setminus \{x_i\}\},$$

which discards all points that have another point in their r-neighborhood. The resulting image measure is called the *type 1 Matérn hard-core point process*. It is stationary if the underlying PPP is stationary. Observe that the intensity of points under the thinning map is reduced more than it has to be in order to obey the hard-core constraint, since always both neighboring points are removed, see ◻ Fig. 2.2 for an illustration.

To construct a thinning that discards less points while still obeying the hard-core constraint, we can use an i.i.d. marked PPP $((X_i, U_i))_{i \in I}$ where the U_i are independent uniform random variables on $[0, 1]$. Then we define the Boolean variables $M_i = \mathbb{1}\{U_i < U_j \text{ for all } X_j \in B_r(X_i) \setminus \{X_i\}\}$, which can be interpreted as marking device X_i with 1 if and only if X_i is the youngest in its coverage zone $B_r(X_i) = \{x \in \mathbb{R}^d \colon |x - X_i| < r\}$. The resulting random point process

$$\sum_{i \in I} M_i \delta_{X_i}$$

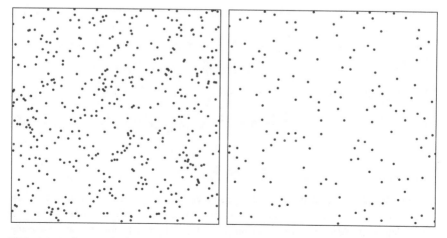

◻ **Fig. 2.2** Realization of a homogeneous PPP on the left and its thinned version on the right, based on the thinning rule of the type 1 Matérn hard-core point process has the desired property and is referred to as the *type 2 Matérn hard-core point process*. Note that, due to the dependence in the thinning procedure, this process is not a PPP, but at least if the underlying PPP is stationary, so is the Matérn hard-core point process.   ◇

Exercise 2.4.10

Calculate the intensities of the type 1 and type 2 Matérn hard-core point processes (see Example 2.4.9) and verify that indeed the type 2 process has a larger intensity.   ◇

Exercise 2.4.11

In the setting of Remark 2.4.6, show that if the kernel K has a shift-invariant Lebesgue density, i.e., $K(x, \mathrm{d}y) = \rho(x - y)\,\mathrm{d}y$ for some density ρ, then a homogenous PPP is equal in distribution to its K-displaced PPP.   ◇

2.5   Palm Calculus

Let Φ be a stationary point process on $\mathbb{R}^d$, that is, the distribution of Φ equals the one of $\Phi + x$ for any $x \in \mathbb{R}^d$. We would like to imagine that we are standing on one of the points $X_i \in \Phi$ of the process and consider the other points from there. In other words, we are interested in the process $\Phi - X_i$, seen from the perspective of X_i. To be sure, this point X_i should be a *typical* one, i.e., not a point that is sampled according to any specific criterion. Since we are in a stationary setting, the randomly chosen point X_i can be assumed to be located at the origin. Hence, as we will explain heuristically below, we would like to consider the conditional version of Φ given $o \in \Phi$. The definition of this object needs some care, since the event $\{o \in \Phi\}$ has probability zero for stationary point processes. The mathematically sound setup for this is *Palm theory*. Let us start by giving the associated existence and uniqueness result including the proof.

Theorem 2.5.1 (Refined Campbell Theorem)
Suppose that Φ is a stationary point process on $\mathbb{S}(\mathbb{R}^d)$ with finite positive intensity λ. Then, there exists a unique probability measure P^o on $\mathbb{S}(\mathbb{R}^d)$ (with corresponding expectation E^o) such that

$$\lambda^{-1}\mathbb{E}\Big[\sum_{i \in I} f(X_i, \Phi - X_i)\Big] = \int_{\mathbb{R}^d} E^o[f(x, \cdot)]\,\mathrm{d}x, \qquad (2.5.1)$$

for all measurable $f \colon \mathbb{R}^d \times \mathbb{S}(\mathbb{R}^d) \to [0, \infty)$. The measure P^o is called the Palm *distribution of Φ.*

Note that if f only depends on x, the refined Campbell theorem is the Campbell Theorem 2.3.1 for stationary processes. Note also that some authors define P^o in such a way that it includes the normalization. It will also be convenient for us to use the unnormalized Palm measure in ▶ Sect. 6.6. Sometimes, it will also be convenient to introduce an $\mathbb{S}(\mathbb{R}^d)$-valued random variable Φ^* with distribution P^o on the original probability space. For example, if f does not depend on x, we then write

$$\lambda^{-1}\mathbb{E}\bigg[\sum_{i\in I}\mathbb{1}\{X_i\in[0,1]^d\}f(\Phi-X_i)\bigg]=\mathbb{E}[f(\Phi^*)]. \tag{2.5.2}$$

Proof

We prove the statement for $f=\mathbb{1}_{B\times A}$ for bounded measurable $B\subset\mathbb{R}^d$, $A\subset\mathbb{S}(\mathbb{R}^d)$. The full statement then follows by the usual monotone class arguments. We define

$$\nu_A(B)=\mathbb{E}\bigg[\sum_{i\in I}\mathbb{1}\{X_i\in B\}\mathbb{1}\{\Phi-X_i\in A\}\bigg].$$

Then, by stationarity,

$$\nu_A(B+x)=\mathbb{E}\bigg[\sum_{i\in I}\mathbb{1}\{X_i-x\in B\}\mathbb{1}\{\Phi-X_i\in A\}\bigg]$$

$$=\mathbb{E}\bigg[\sum_{i\in I}\mathbb{1}\{X_i\in B\}\mathbb{1}\{\Phi-X_i\in A\}\bigg]=\nu_A(B),$$

and thus ν_A is also stationary. Further, since $\nu_A(B)\leq\mathbb{E}[\Phi(B)]=\lambda|B|$, ν_A is also locally finite, and thus ν_A must be equal to $\lambda_A dx$ with $\lambda_A=\nu_A([0,1]^d)$. Then, defining the probability measure $P^o(A)=\lambda_A/\lambda$, we have

$$\mathbb{E}\bigg[\sum_{i\in I}\mathbb{1}\{X_i\in B\}\mathbb{1}\{\Phi-X_i\in A\}\bigg]=\lambda P^o(A)|B|=\lambda\int_{\mathbb{R}^d}\mathbb{E}^o[\mathbb{1}_{A\times B}(x,\cdot)]\,dx.$$

Conversely, for $B\subset\mathbb{R}^d$ with $0<|B|<\infty$, the equality (2.5.1) yields

$$P^o(A)=(\lambda|B|)^{-1}\mathbb{E}\bigg[\sum_{i\in I}\mathbb{1}\{X_i\in B\}\mathbb{1}\{\Phi-X_i\in A\}\bigg], \tag{2.5.3}$$

which shows uniqueness. □

Remark 2.5.2 (Heuristics for Palm Distributions, Typical Points and Conditioning) Let us provide some illustration and interpretation of the Palm distribution. The left-hand side of (2.5.1) can be interpreted as the probability of an event for Φ, seen from a 'typical' device $X_i\in\Phi$, i.e., from a device that is picked uniformly at random. But what does 'uniformly' mean for an infinite point cloud? And what about the right-hand side of (2.5.1)?

To give some substance to the idea of picking a 'typical' point, we pick X_i uniformly at random from the stationary point process Φ with intensity measure λdx for some $\lambda\in(0,\infty)$, in some compact set A with positive Lebesgue measure, say a centered box. That is, we consider the distribution of $\sum_{i\in I}\delta_{X_i}(A)\delta_{\Phi-X_i}$, properly normalized. Note that the normalization is $1/\mathbb{E}[\sum_{i \in I} \delta_{X_i}(A)] = 1/\mathbb{E}[\Phi(A)] = 1/\lambda|A|$. The probability of an event Γ in $\mathbb{S}(\mathbb{R}^d)$ is then equal to

$$\frac{1}{\lambda|A|}\mathbb{E}\Big[\sum_{i \in I} \delta_{X_i}(A)\mathbb{1}\{\Phi - X_i \in \Gamma\}\Big].$$

Actually, it turns out that this does not depend on A. Indeed, considering a partition $(D_k)_{1 \le k \le n}$ of A consisting of connected Lebesgue-positive sets, we can rewrite this as

$$\frac{1}{\lambda|A|}\sum_{k=1}^{n}\mathbb{E}\Big[\sum_{i \in I} \delta_{X_i}(D_k)\mathbb{1}\{\Phi - X_i \in \Gamma\} \,\Big|\, \Phi(D_k) > 0\Big]\mathbb{P}(\Phi(D_k) > 0).$$

Now, considering the limit of finer and finer partitions, we observe that $\mathbb{P}(\Phi(D_k) > 0) = \lambda|D_k| + o(|D_k|)$, and thus the above sum, in the spirit of a Riemann sum, should converge to a limiting expression of the form

$$\frac{1}{\lambda|A|}\lambda\int_A \mathbb{P}\big(\Phi - x \in \Gamma \mid x \in \Phi\big)\,\mathrm{d}x, \tag{2.5.4}$$

ignoring that the conditioning is on a null-set. By translation invariance, the integrand should be $\mathbb{P}(\Phi - x \in \Gamma \mid x \in \Phi) = \mathbb{P}(\Phi \in \Gamma \mid o \in \Phi)$, where we write o for the origin, i.e., it does not depend on x. We arrive at the formal equality

$$\frac{1}{\mathbb{E}[\Phi(A)]}\mathbb{E}\Big[\sum_{i \in I} \delta_{X_i}(A)\mathbb{1}\{\Phi - X_i \in \Gamma\}\Big] = \mathbb{P}(\Phi \in \Gamma \mid o \in \Phi).$$

Note that the right-hand side is independent of A. This equality explains heuristically the relationship between the idea of a PPP seen from a typical point and the process conditioned on having a point at the origin. The distribution $\mathbb{P}(\Phi \in \cdot \mid o \in \Phi) = P^o(\cdot) = \mathbb{P}(\Phi^* \in \cdot)$ is then the Palm version of the distribution of Φ. ◇

Remark 2.5.3 (Reduced Palm Distribution)

1. *Reduced Campbell–Little–Mecke formula.* Following the line of ideas in Remark 2.5.2, a closely related result can be formulated, called the *reduced Campbell–Little–Mecke formula*, which does not require stationarity. It states the existence of the *reduced Palm distribution* $P_x^!$ for $x \in \mathbb{R}^d$, characterized by the equation

$$\mathbb{E}\Big[\sum_{i \in I} f(X_i, \Phi - \delta_{X_i})\Big] = \int E_x^![f(x, \cdot)]\,\mu(\mathrm{d}x), \tag{2.5.5}$$

where μ is the intensity measure of Φ. Note that here the process is not shifted but rather a random point is removed, compare also to the expression (2.5.4). Based on the same heuristic as in the previous remark, the reduced Palm distribution can be interpreted as the distribution of the point process, conditioned on having a point at x that is not evaluated, that is, $P_x^!(\Gamma) = \mathbb{P}(\Phi - \delta_x \in \Gamma \mid x \in \Phi)$. In the context of wireless communication networks, we can use $P_x^!$, for example, to fix a device at position x and analyze the remaining devices because only they cause interference towards x.

2. *Non-reduced Palm distribution.* Based on the definition of the reduced Palm distribution $P_x^!$, a *non-reduced Palm distribution* can also be defined via $P_x(\Gamma) = P_x^!(\{\phi : \phi + \delta_x \in \Gamma\})$ for any $\Gamma \subset \mathbb{S}(\mathbb{R}^d)$ and $x \in \mathbb{R}^d$. Again formally, we can write $P_x(\Gamma) = \mathbb{P}(\Phi \in \Gamma \mid x \in \Phi)$. If the underlying point process is stationary, then the reduced as well as the non-reduced Palm distributions are independent of x. In particular, the non-reduced Palm distribution and the Palm distribution as defined in (2.5.1) satisfy the relation

$$P_x(\Gamma) = P^o(\Gamma + x), \qquad x \in \mathbb{R}^d,$$

in particular $P_o(\Gamma) = P^o(\Gamma)$. ◇

Exercise 2.5.4

Prove the reduced Campbell–Little–Mecke formula by showing that, for any $\Gamma \subset \mathbb{S}(\mathbb{R}^d)$, the *reduced Campbell measure* on $\mathbb{S}(\mathbb{R}^d)$ given by

$$A \mapsto C^!(A \times \Gamma) = \mathbb{E}\Big[\sum_{i \in I} \delta_{X_i}(A) \mathbb{1}\{\Phi - \delta_{X_i} \in \Gamma\}\Big],$$

is absolutely continuous with respect to μ. The resulting density, which depends on Γ, can then be chosen as a probability measure on $\mathbb{S}(\mathbb{R}^d)$, giving rise to the reduced Palm distribution. ◇

Exercise 2.5.5

Let Γ be a measurable subset of $\mathbb{S}(\mathbb{R}^d)$. Prove that, for any stationary point process Φ with finite positive intensity, the following two statements are equivalent. $P^o(\Gamma) = 1$ and $\mathbb{P}$-almost surely $\#\{i \in I : \Phi - X_i \notin \Gamma\} = 0$. *Hint:* Use (2.5.3). ◇

Exercise 2.5.6

Show that if (the distribution of) the random point cloud Φ is isotropic (i.e., rotationally invariant) and translation invariant, then P^o is isotropic. ◇

The refined Campbell Theorem 2.5.1, is formulated for general stationary point processes. In the special case of a homogeneous PPP, the Palm distribution has a particularly simple form, as can be understood from the following argument. Since the event $\{o \in \Phi\}$ has probability zero under the PPP Φ, we instead condition, for some $\varepsilon > 0$, on the event $\{\Phi(B_\varepsilon(o)) = 1\}$ (which has positive probability) and perform the limit $\varepsilon \downarrow 0$. Let us see what this gives for a counting variable $\Phi(A)$ for some bounded open set $A \subset \mathbb{R}^d$ containing the origin o. For any $n \in \mathbb{N}_0$,

$$\mathbb{P}\big(\Phi(A) = n \mid \Phi(B_\varepsilon(o)) = 1\big) = \frac{\mathbb{P}\big(\Phi(A \setminus B_\varepsilon(o)) = n - 1, \Phi(B_\varepsilon(o)) = 1\big)}{\mathbb{P}\big(\Phi(B_\varepsilon(o)) = 1\big)}$$

$$= \mathbb{P}\big(\Phi(A \setminus B_\varepsilon(o)) = n - 1\big)$$

which converges to $\mathbb{P}(\Phi(A \setminus \{o\}) = n - 1) = \mathbb{P}((\Phi + \delta_o)(A) = n)$ as $\varepsilon \downarrow 0$. This suggests that the limiting conditioned process should be nothing but the process $\Phi + \delta_o$. This is made precise in the following result. It states that PPPs are even characterized by this property.

Theorem 2.5.7 (Stationary Mecke–Slivnyak Theorem)
Let Φ be a stationary point process with intensity $\lambda > 0$ and Palm distribution P^o. Then, Φ is a PPP if and only if for all measurable $\Gamma \subset \mathbb{S}(D)$ we have that

$$P^o(\Gamma) = \mathbb{P}(\Phi + \delta_o \in \Gamma). \tag{2.5.6}$$

Proof
First, let Φ be a stationary PPP. It suffices to verify (2.5.6) for events Γ of the form $\Gamma = \{\Phi(A) = n\}$ for open sets $A \subset \mathbb{R}^d$ and $n \in \mathbb{N}_0$. Then, by the refined Campbell Theorem 2.5.1 applied to $f_\varepsilon(x, \Phi) = \mathbb{1}\{x \in B_\varepsilon(o)\}\mathbb{1}\{\Phi(A) = n\}$, we have that

$$P^o(\Gamma) = (|B_\varepsilon(o)|\lambda)^{-1}\mathbb{E}\Big[\sum_{i \in I} \mathbb{1}\{X_i \in B_\varepsilon(o)\}\mathbb{1}\{\Phi(A + X_i) = n\}\Big], \qquad \varepsilon > 0.$$

If $o \notin A$, then there exists an $\varepsilon > 0$ such that also $B_{2\varepsilon}(o) \subset \mathbb{R}^d \setminus A$. In this case, by the independence of the PPP and stationarity, we can conclude

$$P^o(\Gamma) = \mathbb{P}(\Phi(A) = n) = \mathbb{P}(\Phi + \delta_o \in \Gamma).$$

On the other hand, if $o \in A$, for sufficiently small $\varepsilon > 0$, we have $B_\varepsilon(o) \subset A$. Further, we can condition on the event $\{\Phi(B_\varepsilon(o)) = 1\}$ since all other probabilities tend to zero as ε tends to zero. In particular, again by the independence,

$$P^o(\Gamma) = \lim_{\varepsilon \downarrow 0} \mathbb{P}\big(\Phi(A \setminus B_\varepsilon(o)) = n - 1\big) = \mathbb{P}(\Phi + \delta_o \in \Gamma).$$

For the other direction, note that $P^o(\Gamma) = P^!_o(\{\phi \colon \phi + \delta_o \in \Gamma\})$. Using the reduced Campbell–Little–Mecke formula (2.5.5) with $f(x, \Phi) = \mathbb{1}\{x \in A\}\mathbb{1}\{\Phi(A) = n\}$ and $\Gamma = \{\Phi(A) = n\}$, we have on the one hand that

$$\lambda \int E^!_x[f(x, \cdot)]\, dx = \mathbb{E}\Big[\sum_{i \in I} \mathbb{1}\{x \in A\}\mathbb{1}\{(\Phi - \delta_{X_i})(A) = n\}\Big]$$

$$= (n + 1)\mathbb{P}(\Phi(A) = n + 1).$$

On the other hand, using the assumption (2.5.6), we have

$$\lambda \int E^!_x[f(x, \cdot)]dx = \lambda|A|\mathbb{P}(\Phi(A) = n).$$

But this implies that

$$\mathbb{P}(\Phi(A) = n) = \mathbb{P}(\Phi(A) = 0)\frac{(\lambda|A|)^n}{n!}.$$

Using the first characterization of a PPP presented in Exercise 2.2.10, this concludes the proof. □

In view of the representation (2.5.2), the stationary Mecke–Slivnyak theorem can be reformulated with respect to the random variable Φ^*. It then states that a stationary PPP is characterized by the fact that

$$\mathbb{E}[f(\Phi^*)] = \mathbb{E}[f(\Phi + \delta_o)], \qquad f : \mathbb{S}(\mathbb{R}^d) \to [0, \infty) \text{ measurable.}$$

Remark 2.5.8 (Mecke–Slivnyak Theorem) The stationary Mecke–Slivnyak theorem is a special case of the more general *Mecke–Slivnyak theorem*, which does not use stationarity but some mild assumptions on the intensity measure μ. It states that a PPP Φ is characterized by the equation

$$\mathbb{E}\left[\sum_{i \in I} f(X_i, \Phi)\right] = \int \mathbb{E}[f(x, \Phi + \delta_x)]\, \mu(dx),$$

$$f : \mathbb{R}^d \times \mathbb{S}(\mathbb{R}^d) \to [0, \infty) \text{ mb.} \tag{2.5.7}$$

In other words, for Φ a PPP, the reduced Palm distribution $P_x^!$ is equal to the original distribution for μ-almost all x.

The Mecke–Slivnyak theorem can also be seen as a generalization of Campbell's Theorem 2.3.1, which considers functions $f(X_i, \Phi) = f(X_i)$ not depending on Φ. ◊

Exercise 2.5.9 (Subducting Points from PPPs)
Formulate and prove that, for any $n \in \mathbb{N}$ and for any $x_1, \ldots, x_n \in \mathbb{R}^d$, the PPP Φ has the same distribution as the conditional process $\Phi \setminus \{x_1, \ldots, x_n\}$ given that $x_1, \ldots, x_n$ belong to Φ. ◊

Example 2.5.10 (Contact Distance Distribution for Homogeneous PPPs)
Recall from Remark 2.2.12 the contact distance $\text{dist}(x, \phi)$ of a particle $x \in \phi$ to a point set $\phi = \{x_i : i \in I\} \in \mathbb{S}(D)$ without x. If Φ is a homogeneous PPP with intensity λ, then $\mathbb{P}(\text{dist}(o, \Phi^*) \leq r) = \mathbb{P}(\text{dist}(o, \Phi) \leq r) = \mathbb{P}(\text{dist}(x, \Phi) \leq r)$ for any $x \in \mathbb{R}^d$. In words, the distance of a typical point from a homogeneous PPP, positioned at the origin, to its nearest neighbor in the PPP is distributed exactly as the distance from any fixed point. ◊

Exercise 2.5.11
Show that the distribution of a not necessarily stationary PPP with intensity measure μ is equal to its reduced Palm distribution, see Remark 2.5.3, for μ-almost all points. ◊

Exercise 2.5.12

Consider a stationary PPP on $\mathbb{R}^d$ with intensity $\lambda > 0$. For a typical point $X_i \in \Phi$, what is the distribution of the number of points with distance at most $r > 0$ from X_i? ◇

Exercise 2.5.13

Calculate the intensity of the type 1 Matérn hard-core point process from Example 2.4.9 using the stationary Mecke–Slivnyak theorem (2.5.7). ◇

Random Environments: Cox Point Processes

Modeling a system of telecommunication devices in space via a homogeneous PPP represents a situation where no information about the environment or any preferred behavior of the devices is available. To some degree this can be compensated by the use of a non-homogeneous PPP with general intensity measure μ, where areas can be equipped with higher or lower device density. Thereby we leave the mathematically nicer setting of spatial stationarity, but at least we keep the spatial independence. Nevertheless, the independence of devices is an assumption that is often violated in the real world since user behavior is usually correlated. One way to incorporate dependencies into the distribution of devices in space is to use *Cox point processes*, which are PPPs with a random intensity measure representing the environment. In the simplest case, the scalar intensity $\lambda > 0$ of a homogeneous PPP can now be seen as a random variable Λ taking values in $[0, \infty)$ representing, for example temporal fluctuations in the intensity of devices in a city. The joint distribution remains stationary, but it is highly spatially correlated and in particular not ergodic, see ▶ Chap. 6. Cox point processes still belong to the standard examples of point processes and further material can be found, for example, in the textbooks [Hän12, DalVer03, DalVer08, CKMS13].

In the first part of this chapter, ▶ Sect. 3.1, we will present further examples of random environments, relevant for the telecommunication application. They are modeled via random intensity measures Λ for which we exhibit an important criterion called *stabilization* that helps us to quantify the degree of spatial dependence in Λ. In ▶ Sect. 3.2, we will feature one particular class of examples for random environments, *random tessellations*. Random tessellations form a large subfield of stochastic geometry in its own right with a number of textbooks solely dedicated to such processes, for instance [OBSC00, MølSto07, Møl12]. The most prominent example here is the *Poisson–Voronoi tessellation* (PVT), where every Poisson point constitutes the center of a cell that contains all nearest neighbor points in $\mathbb{R}^d$. In the context of telecommunications, this tessellation can be used to model, for example serving zones of base stations, but the collection of cell boundaries also resembles urban street systems. Moreover, the PVT turns out to be precisely the structure needed to recover the distribution of a point process from its Palm version this is the content of ▶ Sect. 3.3.

© Springer Nature Switzerland AG 2020
B. Jahnel, W. König, *Probabilistic Methods in Telecommunications*, Compact Textbooks in Mathematics,
https://doi.org/10.1007/978-3-030-36090-0_3

Throughout the chapter, we present a selection of random tessellation processes with relevance to telecommunication networks and derive first properties.

3.1   Cox Point Processes

In simple words, a *Cox point process* (CPP) is a PPP with random intensity measure Λ. The *directing random measure* Λ can be interpreted as a *random environment* and the resulting processes is thus constructed via a two-level stochastic procedure. More specifically, let $D \subset \mathbb{R}^d$ be a measurable set, and we assume Λ to be a random element of the space of all σ-finite measures $\mathcal{M}(D)$ on D equipped with the smallest sigma algebra such that all evaluation mappings $\mu \mapsto \mu(B)$ from $\mathcal{M}(D)$ to $[0, \infty]$ are measurable for all measurable $B \subset D$. We call such Λ a *random measure* on D. We have the following definition.

Definition 3.1.1 (Cox Point Process)

Let Λ be a random measure on D, then the PPP Φ with random intensity measure Λ is called a *Cox point process directed by* Λ.

For an illustration of a realization of a CPP, see ◻ Fig. 3.1. Let us make some first comments.

Remark 3.1.2   (Properties of CPPs)
1. *Expected number of points.* For a CPP, the expected number of points in a measurable volume $A \subset D$ is given by the expected intensity of A, i.e.,

$$\mathbb{E}[\Phi(A)] = \mathbb{E}\big[\mathbb{E}[\Phi(A)|\Lambda]\big] = \mathbb{E}[\Lambda(A)].$$

2. *Laplace transform.* The Laplace transform of a CPP is given by

$$\mathcal{L}_\Phi(f) = \mathbb{E}[e^{-\Phi(f)}] = \mathbb{E}\Big[\exp\Big(\int_D (e^{-f(x)} - 1)\,\Lambda(dx)\Big)\Big], \tag{3.1.1}$$

for all measurable $f : D \to [0, \infty)$.   ◇

The theory of CPPs provides a broad setting for modeling interesting spatial telecommunication systems.

Example 3.1.3   (Absolutely Continuous Random Fields and Hot Spots)
A large class of random environments Λ is given by measures having a non-negative random field $\ell = \{\ell_x\}_{x \in \mathbb{R}^d}$ as a Lebesgue density, i.e., $\Lambda(dx) = \ell_x\,dx$. On $\mathbb{R}^d$, one often assumes ℓ to be stationary. The simplest example is the *mixed PPP* where $\Lambda(dx) = Z dx$ and Z a non-negative random variable. Other examples can be constructed via *measures modulated by a random closed set* Ξ, see for example [CKMS13, Section 5.2.2]. Here, $\ell_x = \lambda_1 \mathbb{1}\{x \in$

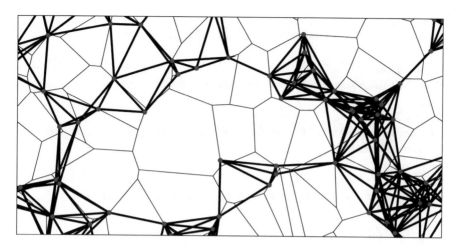

■ **Fig. 3.1** Realization of the Gilbert graph, see ▶ Chap. 4, of devices confined to a street system given by a Poisson–Voronoi tessellation

$\Xi\} + \lambda_2 \mathbb{1}\{x \notin \Xi\}$ with parameters $\lambda_1, \lambda_2 \geq 0$. For instance, Ξ could be given by the Boolean model $\bigcup_{j \in J} B_r(Y_j)$ of a PPP $(Y_j)_{j \in J}$, see ▶ Chap. 4, interpreted as a random configuration of hot spots. Another important example is a random measure induced by a *shot-noise field*, see [CKMS13, Section 5.6]. Here, $\ell_x = \sum_{j \in J} k(x - Y_j)$ for some non-negative integrable kernel $k \colon \mathbb{R}^d \to [0, \infty)$ with compact support and an independent PPP $(Y_j)_{j \in J}$. ◊

Example 3.1.4 (Singular Random Fields and Random Street Systems)

Very interesting for a realistic modeling of an urban area is a random environment Λ that is defined as the restriction of the Lebesgue measure to a random segment process S in $\mathbb{R}^d$. For $d = 2$ one can then, for example, think of a random street system. More precisely, S is a point process in the space of line segments [CKMS13, Chapter 8], which may be stationary, see ■ Fig. 3.1 for an illustration. Observe that S is a union of one-dimensional subsets, in particular a nullset with respect to the two-dimensional Lebesgue measure, and thus singular. There is a natural measure ν_1 on S that assigns a finite and positive value to each bounded non-trivial line segment (indeed, its length). This measure is the one-dimensional Hausdorff measure ν_1 on S. Then we put $\Lambda(\mathrm{d}x) = \nu_1(S \cap \mathrm{d}x)$ and obtain a random measure on $\mathbb{R}^d$ that is concentrated on S.

There are a number of interesting choices of S, some of which have high relevance as models for street systems of cities or rural areas. In the next ▶ Sect. 3.2, we introduce several tessellations and more general random graphs, for example the Poisson–Voronoi, Poisson–Delaunay and the Poisson-line tessellations, and highlight some of their properties. ◊

Exercise 3.1.5

In the spirit of (2.5.3), define the Palm distribution of a stationary CPP.   ◊

Exercise 3.1.6

Calculate the nearest neighbor distance distribution for the mixed PPP.   ◊

Exercise 3.1.7

Calculate the intensity of a CPP with directing measure given by the shot-noise field based on a homogeneous PPP.   ◊

Exercise 3.1.8

Let Φ be a stationary CPP with intensity $\lambda \Lambda$. Like in Exercise 2.3.3, prove that

$$\frac{d}{d\lambda}\mathbb{E}[\Phi(f)] = \mathbb{E}\left[\int f(x)\,\Lambda(dx)\right]$$

for all Lebesgue integrable $f: \mathbb{R}^d \to \mathbb{R}$.   ◊

3.2   Random Tessellations

Prominent examples of singular random environments as in Example 3.1.4 are given by directing random measures Λ derived from some *tessellation* S via $\Lambda(dx) = \nu_1(S \cap dx)$. These random tessellations are also of independent interest and can be used in a number of seemingly unrelated branches of mathematics, such as numerical methods for partial differential equations or algorithmic geometry. General accounts on random tessellations can be found, for example, in [OBSC00, MølSto07, Møl12, vLie12].

The most common tessellation is the *Poisson–Voronoi tessellation* (PVT), which we introduce now. Consider a PPP $\Phi = (X_i)_{i \in I}$ in D. We assign to each $X_i \in \Phi$ the *Poisson–Voronoi cell*

$$z(X_i) = z(X_i, \Phi) = \left\{x \in D: |x - X_i| \leq \inf_{j \in I} |x - X_j|\right\}. \tag{3.2.1}$$

In words, the interior of $z(X_i)$ contains all points in D that are closer to X_i than to any other point in Φ. Now, D is partitioned into cell interiors and cell boundaries, which motivates the term tessellation. It can be proved that the cell boundaries are polygon lines and the cells are convex almost surely. See Fig. 3.2 on the left for an illustration. If the underlying PPP Φ is homogeneous, then the distribution of the PVT is translation invariant and isotropic, i.e., invariant with respect to rotations around the origin. A number of important characteristics of the PVT, such as the expected cell volume, cell perimeters, etc., can be calculated from the intensity of Φ, see [OBSC00, Table 5.1.1].

In telecommunication applications, one can, for example see X_i as the location of a base station and $z(X_i)$ as its serving zone, but this is not the interpretation that we are after here. Instead, we interpret S as a random street system, and indeed there is some

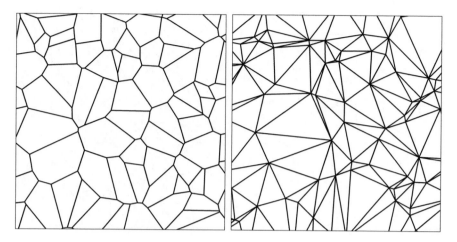

□ Fig. 3.2 Realizations of the PVT (left) and the PDT (right)

statistical evidence indicating that PVTs give decent fits to street systems in central European cities, see [Cou12].

Let us mention a few more examples of tessellation processes. First, the *Poisson–Delaunay tessellation (PDT)* is the dual tessellation corresponding to the PVT. Here line segments are drawn such that any two points are connected by a line if and only if their Poisson–Voronoi cell boundaries have a non-empty intersection. The PDT naturally has very similar locality properties as the PVT, but completely different behavior, for example, of its vertex degree. More precisely, the typical Poisson–Voronoi cell boundary consists of 6 line segments leading to an expected degree of 6 for the typical Poisson–Delaunay vertex. But, in the PVT, with probability one, only 3 line segments meet in a vertex. See □ Fig. 3.2 on the right for an illustration.

The *Poisson–Gabriel graph* (PGG) is a subgraph of the PDT where only those Delaunay edges are kept, which directly cross the associated Voronoi edge. It is also a tessellation, and an equivalent construction can be defined as follows. We draw a line between any two Poisson points and the disc having this line as a diagonal. The line is a Gabriel edge if and only if the associated disc contains no other Poisson points.

Another subgraph of the PDT is the *Poisson-relative-neighborhood graph* (PRNG), where any two points X_i, X_j in the underlying PPP Φ are connected by an edge if and only if for all other points $X_k \in \Phi \setminus \{X_i, X_j\}$ we have that $|X_k - X_i| \vee |X_k - X_j| > |X_i - X_j|$. The appeal for considering the PRNG in the context of telecommunications comes from the fact that it features dead-ends. By this we mean that when we use random graphs as models for street systems, then especially in rural areas, dead-end roads can no longer be neglected and the PRNG is one possible model for this scenario. See □ Fig. 3.3 on the left for an illustration. The PGG and the PRNG are part of a one-parameter family of tessellations called β-skeletons.

The *Poisson-line tessellation* (PLT) is a tessellation based on a PPP on $\mathbb{R} \times [0, \pi)$ or equivalently a PPP on $\mathbb{R}$ with i.i.d marks in $[0, \pi)$ representing random angles. For every

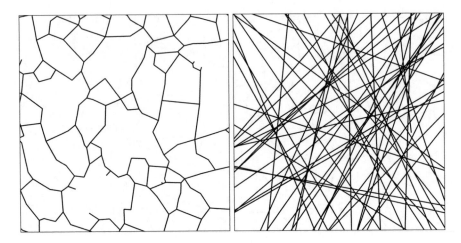

Fig. 3.3 Realizations of the PRNG (left) and the PLT (right)

Poisson point (p_i, α_i), we then define the line $\{(x, y) \in \mathbb{R}^2 : x \cos \alpha_i + y \sin \alpha_i = p_i\}$, see ■ Fig. 3.3 on the right for an illustration.

As yet another example of a tessellation process with relevance to telecommunications, let us mention *Manhattan grids* (MG), with the particular example of the *rectangular Poisson-line process* (RPLT). The RPLT consists of perpendicular lines through the points of independent PPPs representing landmarks along each axis in $\mathbb{R}^2$. Despite its popularity in stochastic geometry, this model has the serious drawback of only being able to represent street systems where the distance between successive streets is exponentially distributed. This constraint can be removed by replacing the PPP by a *stationary renewal process*, that is, by a renewal process that is statistically invariant under shifts along the axis. See ■ Fig. 3.4 on the left for an illustration. The MG can be further refined by putting additional rectangular lines inside the boxes given by the MG. This construction gives rise to *nested Manhattan grids* (NMG), see ■ Fig. 3.4 on the right for an illustration.

Exercise 3.2.1
Calculate the street intensity $\mathbb{E}[\nu_1(S \cap [0, 1]^2)]$ where S is your favorite tessellation process from above.   ◇

Exercise 3.2.2
Consider the PVT based on a PPP with intensity $\lambda > 0$ and denote its distribution by $\mathbb{P}_\lambda$. Verify the following scaling relation. For all $a > 0$ we have that $\mathbb{E}_\lambda[\nu_1(S \cap [0, 1]^2)] = a^{-1}\mathbb{E}_{a^{-2}\lambda}[\nu_1(S \cap [0, a]^2)]$.   ◇

Exercise 3.2.3
Consider the PVT based on the homogeneous PPP on $\mathbb{R}^2$ with intensity $\lambda > 0$. Verify the following properties. The area of the typical cell is $1/\lambda$. The expected number of vertices and

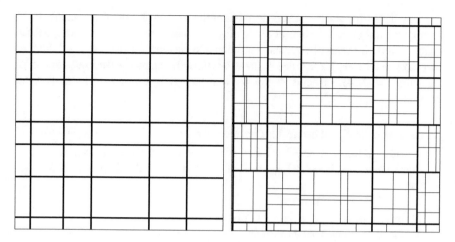

□ Fig. 3.4 Realizations of the MG (left) and the NMG (right)

the expected number of edges of a typical cell boundary is 6. The expected perimeter of the typical cell is $4/2\sqrt{\lambda}$ and the expected length of a typical edge is $2\sqrt{\lambda}/(3\lambda)$.   ◊

So far we have presented tessellations that can be based on PPPs. Let us at least mention that an even richer family of random tessellations can be defined via rules that do not exclusively depend on spatial positions and furthermore on additional random information. In other words, these are tessellations based on marked point processes. One prominent example is the *Johnson–Mehl tessellation* (JMT), see for example [BolRio08]. For this we consider the homogeneous PPP $\Phi = \{(X_i, T_i)\}_{i\in I}$ on $\mathbb{R}^2 \times [0, \infty)$ with intensity measure λLeb and define the Johnson–Mehl metric via

$$d\big((x, s), (y, t)\big) = |x - y| + |t - s|.$$

Then, the JMT is given by

$$\bigcup_{i\in I} \partial\big\{x \in \mathbb{R}^2 : d\big((x, 0), (X_i, T_i)\big) = \inf_{j\in I} d\big((x, 0), (X_j, T_j)\big)\big\},$$

where for a set A, we write $\partial A = \bar{A} \setminus A^o$ for the boundary of A. To give some intuition, as a generalization of the PVT, the Johnson–Mehl cells contain all those points that are close to its center in a spatial dimension combined with a desire to connect to the 'youngest' point, where we interpret the additional dimension $[0, \infty)$ as time. Further tessellations that are based on additional marks are, for example the *Laguerre tessellation* (see [LauZuy08]) or the additively or multiplicatively weighted PVT (see [OBSC00]).

Often, a key to the mathematical analysis of CPPs with random intensity measures of the form $\Lambda(\mathrm{d}x) = \nu_1(S \cap \mathrm{d}x)$ where S is a tessellation process is to understand the mixing properties of Λ. In other words, to understand how strong and how far reaching the spatial stochastic dependencies of the process are. In the remainder of this section, we introduce the concept of *stability* as a tool to measure these dependencies quantitatively. We denote by Λ_B the restriction of a measure Λ to a set $B \subset \mathbb{R}^d$. Further, let $Q_r(x) = x + [-r/2, r/2]^d$ denote the cube with side length $r > 0$ centered at $x \in \mathbb{R}^d$ and put $Q_r = Q_r(o)$. We define $\mathrm{dist}(\varphi, \psi) = \inf\{|x - y| : x \in \varphi, y \in \psi\}$ for the distance between sets $\varphi, \psi \subset \mathbb{R}^d$.

Definition 3.2.4 (Stabilizing Random Measure)

A random measure Λ on $\mathbb{R}^d$ is called *stabilizing* if there exists a random field of *stabilization radii* $\mathcal{R} = \{R_x\}_{x \in \mathbb{R}^d}$ defined on the same probability space as Λ, such that

1. $(\Lambda, \mathcal{R})$ is jointly stationary,
2. $\lim_{r \uparrow \infty} \mathbb{P}(\sup_{y \in Q_r \cap \mathbb{Q}^d} R_y < r) = 1$, and
3. for all $r \geq 1$, the random variables

$$\left\{ f(\Lambda_{Q_r(x)}) \mathbb{1}\{ \sup_{y \in Q_r(x) \cap \mathbb{Q}^d} R_y < r \} \right\}_{x \in \varphi}$$

are independent for all bounded measurable functions $f : \mathcal{M}(D) \to [0, \infty)$ and all finite $\varphi \subset \mathbb{R}^d$ with $\mathrm{dist}(x, \varphi \setminus \{x\}) > 3r$ for all $x \in \varphi$.

A strong form of stabilization is given if Λ is *b-dependent* in the sense that Λ_A and Λ_B are independent whenever $\mathrm{dist}(A, B) = \inf\{|x - y| : x \in A, y \in B\} > b$. The two models of Example 3.1.3 are *b*-dependent for some *b*. The random measure Λ concentrated on the Poisson–Voronoi tessellation S of Example 3.1.4 is *exponentially stabilizing* in the sense that it is stabilizing for some stabilization radii such that $\mathbb{P}(\sup_{y \in Q_r \cap \mathbb{Q}^d} R_y \geq r) \leq \exp(-cr)$ for some constant $c > 0$.

Lemma 3.2.5 (PVTs Are Exponentially Stabilizing) *The random measure* $\Lambda(dx) = \nu_1(S \cap dx)$ *where S is the stationary PVT on $\mathbb{R}^d$ is exponentially stabilizing.*

Idea of Proof
The proof rests on the definition of the radius of stabilization as $R_x = \inf\{|X_i - x| : X_i \in \Phi\}$, for details see [CHJ19]. □

To give some further intuition for the definition of stabilization consider the PVT. There, the defining process of cell centers is a PPP which features complete independence over space. Nevertheless, the deterministic construction of the associated PVT creates potentially far-reaching correlations, essentially due to the following effect. If the underlying PPP has a large void space, then, in this area, the edge set in S can depend on spatial perturbations of Poisson points which are far away. But such a scenario is exponentially unlikely, since large void spaces are exponentially unlikely, see Example 2.5.10 about the contact distance. In order to quantify this, in the definition, the notion of stabilization radii is introduced.

Exercise 3.2.6

Verify that the random environment given by random closed sets Ξ as presented in Exercise 3.1.3, where Ξ is a Boolean model, is b-dependent. ◊

Exercise 3.2.7

Show that the random measure $\Lambda(\mathrm{d}x) = \nu_1(S \cap \mathrm{d}x)$ where S is the PDT is exponentially stabilizing. In this case, to define the stabilization radii as the closest Poisson points will turn out to be insufficient. One possible definition would be to introduce for any x a fixed radial segmentation centered at x and define R_x as the smallest radius such that there are Poisson points in all segments. ◊

Exercise 3.2.8

Show that the mixed PPP of Example 3.1.3 is not stabilizing. ◊

Remark 3.2.9 (Non-stabilizing Random Environments) Note that the RPLT and in particular the MG and NMG are not stabilizing. This comes from the fact that individual edges reach to infinity and thus create infinitely long correlations. ◊

In the context of large deviations, as presented in ▶ Chap. 7, it is sometimes desirable to know that exponential moments of the street length per unit area exist. We present here the associated result.

Theorem 3.2.10 (Exponential Moments)

Let S be the PVT, PDT or PLT based on the homogeneous PPP with intensity $\lambda > 0$. Then, for all $\alpha \in \mathbb{R}$ we have that $\mathbb{E}[\exp(\alpha \nu_1(S \cap [0, 1]^2))] < \infty$.

Idea of Proof

The proof rests on the stabilization property for PVT and PDT, for details see [JahTob19]. □

Exercise 3.2.11

Using the construction of the PLT, show that Theorem 3.2.10 holds. ◊

3.3 Inversion Formulas of Palm Calculus

Recall that the refined Campbell Theorem 2.5.1 defines the Palm distribution P^o for a stationary point process Φ. In Remark 2.5.2 we give a heuristic for the interpretation of P^o as the distribution of Φ conditioned to have a point at the origin. This allows us to properly address questions concerning a typical point. The main result in this section states how the stationary distribution can be retrieved from the Palm distribution via the typical cell of the PVT. Recall the definition of the Voronoi cell $z(X_i, \Phi)$ of a point X_i

in a point cloud Φ given in (3.2.1). As stated in Exercise 2.1.13, for a non-trivial simple stationary point process Φ, there is a unique point $q(o, \Phi)$ that is closest to the origin.

Theorem 3.3.1 (Inversion Formula for Palm Calculus)
Suppose that Φ is a simple stationary point process on $\mathbb{S}(\mathbb{R}^d)$ with intensity $\lambda \in (0, \infty)$. Then, we have that

$$\mathbb{E}[g(q(o, \Phi), \Phi)] = \lambda E^o\left[\int \mathbb{1}\{x \in z(o, \cdot)\} g(-x, \cdot - x) \, dx\right], \tag{3.3.1}$$

for all measurable $g \colon \mathbb{R}^d \times \mathbb{S}(\mathbb{R}^d) \to [0, \infty)$.

Proof
Setting $h(X_i, \Phi) = f(X_i, \Phi - X_i)$ in (2.5.1) in the refined Campbell theorem, we have the reformulation

$$\mathbb{E}\left[\sum_{i \in I} h(X_i, \Phi)\right] = \lambda \int E^o[h(-x, \cdot - x)] \, dx, \tag{3.3.2}$$

where we additionally changed the variable x to $-x$. Now, define $h(X_i, \Phi) = g(X_i, \Phi) \mathbb{1}\{q(o, \Phi) = X_i\}$, then

$$\mathbb{E}\left[\sum_{i \in I} h(X_i, \Phi)\right] = \mathbb{E}[g(q(o, \Phi), \Phi)],$$

and thus the left-hand sides of (3.3.2) and (3.3.1) coincide. On the other hand, we have P^o-almost surely that $q(o, \cdot - x) = -x$ if and only if $x \in C(o, \cdot)$. But this implies that

$$\int E^o[h(-x, \cdot - x)] \, dx = E^o\left[\int \mathbb{1}\{x \in z(o, \cdot)\} g(-x, \cdot - x) \, dx\right],$$

and thus also the right-hand sides of (3.3.2) and (3.3.1) coincide. $\square$

Remark 3.3.2 (Reformulations and Applications of the Inversion Formula) Let us present some reformulations and applications of the inversion formula for Palm calculus.
1. *Coupled versions.* If we denote by Φ^* the random point process Φ under the Palm distribution, realized on a common probability space, then (3.3.1) takes the convenient form

$$\mathbb{E}[g(q(o, \Phi), \Phi)] = \lambda \mathbb{E}\left[\int \mathbb{1}\{x \in z(o, \Phi^*)\} g(-x, \Phi^* - x) \, dx\right].$$

The cell $Z_o = z(o, \Phi^*)$ is referred to as the *typical Voronoi cell* of the point process Φ.

2. *Campbell versions.* For $g(q(o, \Phi), \Phi) = g(\Phi)$ we get

$$\mathbb{E}[g(\Phi)] = \lambda E^o\left[\int \mathbb{1}\{x \in z(o, \cdot)\}g(\cdot - x)\,dx\right], \tag{3.3.3}$$

which characterizes the distribution of Φ by its Palm distribution.

3. *Expected cell volumes.* For $g \equiv 1$ we get the intuitive expression $\mathbb{E}[Z_o] = \lambda^{-1}$. In words, if λ is the expected number of points in the unit volume, then the cell of any point has an expected volume of λ^{-1}.

4. *Expected typical cell volumes.* Taking $g(q(o, \Phi), \Phi) = h(\Phi - q(o, \Phi))$, the inversion formula for Palm calculus yields

$$\mathbb{E}[h(\Phi - q(o, \Phi))] = \lambda \mathbb{E}[|Z_o|h(\Phi^*)],$$

which shows in particular that the distribution of $\Phi - q(o, \Phi)$ is absolutely continuous with respect to the Palm distribution. Moreover, this identity shows that the stationary distribution can be seen as the volume-biased version of its Palm distribution with a random shift. Further denote the Voronoi cell that covers the origin by $C_o(\Phi)$ and set $g(\Phi) = |C_o(\Phi)|^{-1}$, then (3.3.3) yields

$$\mathbb{E}[|C_o(\Phi)|^{-1}] = \lambda = E^o[|Z_o|]^{-1}.$$

Using Jensen's inequality, we thus arrive at the inequality $\mathbb{E}[|C_o|] \le E^o[|Z_o|]$, which can be seen as a spatial version of the *waiting-time paradox*. Indeed, observing the cells that cover the origin C_o under the stationary distribution $\mathbb{P}$, leads to a bias towards seeing larger cells as if the volume of the typical cell Z_o would be observed by the Palm distribution P^o.

Exercise 3.3.3

Under the assumptions of Theorem 3.3.1, show that for all measurable $f \colon \mathbb{R}^d \times \mathbb{S}(\mathbb{R}^d) \to [0, \infty)$ we have that

$$E^o[f(\cdot)] = \lambda^{-1}\mathbb{E}[|C_o(\Phi)|^{-1} f(\Phi - q(o, \Phi))].$$

In words, the Palm distribution can be characterized by a volume-debiased and randomly-shifted version of the stationary distribution. $\Diamond$

Coverage and Connectivity: Boolean Models

In this chapter, we discuss mathematical approaches to the two most fundamental questions about spatial telecommunication models:

- *Coverage:* How much of the area can be reached by the signals emitted from the users, respectively the base stations?
- *Connectivity:* How far can a message travel through the system in a multihop-functionality?

To do this, we introduce and study the most basic model for message transmission within a spatial system formed by a PPP $\Phi = (X_i)_{i \in I}$ of users or base stations, the *Boolean model*. In this model, which we introduce in ▶ Sect. 4.1, to each location X_i a random closed set Ξ_i is attached; Ξ_i is the local communication zone that can be reached by a signal emitted from X_i. Then $\bigcup_{i \in I}(X_i + \Xi_i)$ is the *communication area*, the set of locations that can be reached by any message transmission. In ▶ Sect. 4.2, we study questions about the coverage of a given compact set $C \subset \mathbb{R}^d$, i.e., about the probability that C can be reached by some signal. These are local questions. In contrast, in ▶ Sect. 4.3, we consider global questions about whether or not the communication area possesses an unbounded connected component. This we will do only for homogeneous PPPs and only for balls Ξ_i of a given fixed radius. In this simple but fundamental setting, we will distinguish two drastically different scenarios, the occurrence versus the absence of *percolation*. The distinction is one important example of a *phase transition* and lies at the heart of a beautiful theory called *continuum percolation*. Both phases are non-trivial, as we formulate in ▶ Sect. 4.3. In order to carry out the proof for that in ▶ Sect. 4.5, we first need to develop the discrete counterpart of the theory, which we will do in ▶ Sect. 4.4. Furthermore, we collect in ▶ Sect. 4.6 additional relevant connectivity questions related to percolation, and in ▶ Sect. 4.8 we discuss some peculiarities on percolation that arise for CPPs.

See [BacBla09a] and [BacBla09b] (which we follow in ▶ Sects. 4.1 and 4.2) for applications of the Boolean model to telecommunication, in particular coverage and percolation properties, and see [BolRio06], [MeeRoy96] and [FraMee08] for mathematical proofs of continuum percolation properties. Standard references on the discrete part of the theory are [Gri89] and [BolRio06]. ▶ Section 4.6 is about further

© Springer Nature Switzerland AG 2020
B. Jahnel, W. König, *Probabilistic Methods in Telecommunications*, Compact Textbooks in Mathematics,
https://doi.org/10.1007/978-3-030-36090-0_4

interpretations of the percolation probability in telecommunications and is largely self-contained except for the shape theorem, see [YCG11]. The results about continuum percolation for CPPs presented in ▶ Sect. 4.8 are less standard and taken from [CHJ19].

4.1   Boolean Models

Let $\Phi = (X_i)_{i \in I}$ be a random point process in $\mathbb{R}^d$ with intensity measure μ. Again, we interpret the X_i as the locations of the users or base stations of a spatial telecommunication system. Now, we extend the model by adding a random closed set $\Xi_i \subset \mathbb{R}^d$ around each user X_i and interpret $X_i + \Xi_i$ as the area that can be reached by a signal that is emitted from X_i. We call Ξ_i the *local communication zone* around X_i. The idea is that the strength of the signal decays quickly in the distance, and a certain least strength is necessary for a successful transmission. Typical choices for Ξ_i are centered balls with random or deterministic radii, but more complex choices are thinkable and have their right, e.g., when environmental conditions have to be considered. For example, if X_i is located on a street, then its local communication area Ξ_i will be shaped by the houses left and right of the location X_i and will be approached by some rectangle, depending on the location X_i. Apart from that, we will take the random sets Ξ_i, $i \in I$, as independent. Hence, we would like to see the Ξ_i as marks attached to the users X_i. Note that, under this assumption, the case of Ξ_i being given by a Poisson–Voronoi cell, see (3.2.1), is not covered.

Definition 4.1.1 (Boolean Model)

Let $\Phi = (X_i)_{i \in I}$ be a PPP in $\mathbb{R}^d$, and let K be a probability kernel from $\mathbb{R}^d$ to the set of closed subsets of $\mathbb{R}^d$. Consider the K-marking $\Phi_K = \sum_{i \in I} \delta_{(X_i, \Xi_i)}$ according to Definition 2.4.1. Then, the random set $\Xi_{BM} = \bigcup_{i \in I}(X_i + \Xi_i)$ is called a *Boolean model*.

We have formally taken the set of all closed subsets of $\mathbb{R}^d$ as the mark space, and there is also a natural σ-algebra on this set to turn this into a measurable space. However, this space is not a locally compact topological space, and therefore the above definition, strictly speaking, does not fall into Definition 2.4.1. However, there is no problem concentrating the kernel on a much smaller set of closed sets, e.g., indexed by $\mathbb{R}^l$ for some set of parameters l in a natural way that turns it into a locally compact topological space. One important example is the set of centered balls (or squares, or rectangles, ...) with a random radius. We will only think of such examples in these notes.

For simplicity, from now on we will consider only independent K-markings, i.e., we will assume that the random sets Ξ_i are i.i.d., not depending on the location X_i that they are attached to. That is, the kernel K is just one probability measure, which we will drop from the notation. The Boolean model Ξ_{BM} is interpreted as the total communication area, i.e., the (random) set that is covered by the signals emitted from the set Φ (downlink communications) or alternatively the set of location from which Φ can receive signals (uplink communications).

4.2 Coverage Properties

Let Ξ_{BM} be a Boolean model in the sense of ▶ Sect. 4.1, i.e., a PPP with an independent marking in a locally compact topological mark space. In this section, we provide notions and methods to determine probabilities of *coverage*, i.e., events that a given set or point lies in Ξ_{BM}. That is, we look only at one single transmission step from some X_i. Mathematically, this amounts to the study of the local structure of the random set Ξ_{BM}, i.e., a local question.

By Ξ we denote a generic random closed set that we use in our marking, i.e., a random variable having the distribution K. From now on, we will not use the kernel K anymore, but we will use $\mathbb{P}$ and $\mathbb{E}$ for probability and expectation with respect to Ξ. We will assume that its distribution satisfies

$$\mathbb{E}[\mu(C - \Xi)] < \infty, \qquad \text{for any compact } C \subset \mathbb{R}^d, \tag{4.2.1}$$

where $C - \Xi = \{x - y : x \in C, y \in \Xi\}$. For example, if $C = \{1\}$ and $\Xi = [-R, R]$ with some $R \in (0, \infty)$, then $C - \Xi = [-R+1, 1+R]$ is the set of user locations x such that $x + \Xi$ intersects C. Condition (4.2.1) ensures that the expected number of grains Ξ communicating with any compact C is finite. In particular, under this condition, Ξ_{BM} is also a random closed set.

Exercise 4.2.1
Assume that $\Xi = B_R(o)$ is a ball with random radius R. Show that condition (4.2.1) is equivalent to $\mathbb{E}[R^d] < \infty$. ◇

The *capacity functional* of Ξ is defined as the function

$$T_\Xi(C) = \mathbb{P}(\Xi \cap C \neq \emptyset), \qquad C \subset \mathbb{R}^d \text{ compact.} \tag{4.2.2}$$

This function can be seen as an equivalent of the distribution function of a real random variable. It uniquely determines the distribution of Ξ, according to *Choquet's theorem*, see [Mat75].

Lemma 4.2.2 (Number of Covered Devices) *For any compact set $C \subset \mathbb{R}^d$, the number*

$$\Phi_{BM}(C) = \#\{i \in I : (X_i + \Xi_i) \cap C \neq \emptyset\}$$

is a Poisson random variable with parameter $\mathbb{E}[\mu(C - \Xi)]$.

Proof
Observe that the random point process

$$\sum_{i \in I} \delta_{X_i} \mathbb{1}\{(X_i + \Xi_i) \cap C \neq \emptyset\}$$

is an independent thinning of Φ (recall Lemma 2.2.15) with (space-dependent) thinning probability

$$p_C(x) = \mathbb{P}((x + \Xi) \cap C \neq \emptyset) = \mathbb{P}(x \in C - \Xi).$$

In the same way as in the proof of Lemma 2.2.15, one sees that this process is a PPP with intensity measure $p_C(x)\,\mu(dx)$. Hence, the total number of its points is a Poisson random variable with parameter equal to $\int_{\mathbb{R}^d} p_C(x)\,\mu(dx)$. With the help of Fubini's theorem, we identify this parameter as follows

$$\int p_C(x)\,\mu(dx) = \int \mathbb{P}(x \in C - \Xi)\,\mu(dx) = \mathbb{E}\left[\int \mathbb{1}\{x \in C - \Xi\}\,\mu(dx)\right]$$

$$= \mathbb{E}[\mu(C - \Xi)],$$

which ends the proof.   □

Lemma 4.2.3 (Explicit Expression for the Capacity Functional) *The capacity functional is identified as*

$$T_{\Xi_{BM}}(C) = 1 - e^{-\mathbb{E}[\mu(C-\Xi)]}, \qquad C \subset \mathbb{R}^d \text{ compact.}$$

Proof
Observe that $T_{\Xi_{BM}}(C) = \mathbb{P}(\Phi_{BM}(C) > 0)$ and use Lemma 4.2.2.   □

From now on, we restrict to the stationary (or homogeneous) Boolean model, by which we mean that the intensity measure μ of the underlying PPP is equal to λdx for some $\lambda \in (0, \infty)$, and we call λ the intensity of the Boolean model. It is clear that then also the capacity functional of the Boolean model,

$$T_{\Xi_{BM}}(C) = \mathbb{P}\left(C \cap \bigcup_{i \in I}(X_i + \Xi_i) \neq \emptyset\right), \qquad C \subset \mathbb{R}^d \text{ compact,}$$

(where the probability extends over the PPP Φ and the family of the Ξ_i's) is shift-invariant, i.e., $T_{\Xi_{BM}}(z + C) = T_{\Xi_{BM}}(C)$.

Next, we consider the *volume fraction*

$$p = \frac{\mathbb{E}[|\Xi_{BM} \cap B|]}{|B|} \tag{4.2.3}$$

and note that it does not depend on the compact set $B \subset \mathbb{R}^d$, as long as it has positive Lebesgue measure. The volume fraction has the nice interpretation as the probability that the origin is covered by the Boolean model, as

$$p = \frac{1}{|B|}\int_B \mathbb{E}[\mathbb{1}\{x \in \Xi_{BM}\}]\,dx = \mathbb{E}[\mathbb{1}\{o \in \Xi_{BM}\}] = \mathbb{P}(o \in \Xi_{BM}) = T_{\Xi_{BM}}(\{o\}),$$

where we write $|B|$ for the Lebesgue measure of B. In particular, Lemma 4.2.3 tells us that $p = 1 - \exp(-\lambda \mathbb{E}[|\Xi|])$.

Remark 4.2.4 (Covariance of Coverage Variables) The volume fraction p is the expectation of the *coverage variable* at the origin, $\mathbb{1}\{o \in \Xi_{\mathrm{BM}}\}$. The expectation of the product of the two coverage variables $\mathbb{1}\{o \in \Xi_{\mathrm{BM}}\}$ and $\mathbb{1}\{z \in \Xi_{\mathrm{BM}}\}$ can be calculated in an elementary way as

$$\widetilde{C}(z) = \mathbb{E}\big[\mathbb{1}\{o \in \Xi_{\mathrm{BM}}\}\,\mathbb{1}\{z \in \Xi_{\mathrm{BM}}\}\big]$$
$$= \mathbb{P}(0 \text{ and } z \text{ lie in } \Xi_{\mathrm{BM}}) = 2p - 1 + (1 - p)^2 \exp(-\lambda \mathbb{E}[|\Xi \cap (\Xi + z)|]).$$

The function $\widetilde{C}(z)$ is usually referred to as the *covariance function* of the Boolean model. It is the probability that two points separated by the vector z are covered. The covariance of the two coverage variables is given by

$$C(z) = \widetilde{C}(z) - p^2 = -(1 - p)^2\big(1 - \exp(-\lambda \mathbb{E}[|\Xi \cap (\Xi + z)|])\big). \qquad \Diamond$$

Exercise 4.2.5

Consider the stationary Boolean model with intensity λ. Show that for any measurable set C, the measure of the vacancy region, $|C \setminus \Xi_{\mathrm{BM}}|$, has expectation given by

$$\mathbb{E}[|C \setminus \Xi_{\mathrm{BM}}|] = |C| \exp(-\lambda \mathbb{E}[|\Xi|])$$

and second moment given by

$$\mathbb{E}[|C \setminus \Xi_{\mathrm{BM}}|^2] = \int_{C \times C} \exp\big(-2\lambda\mathbb{E}[|\Xi|] + \lambda\mathbb{E}[|(x - y + \Xi) \cap \Xi|]\big)\,\mathrm{d}x\mathrm{d}y. \qquad \Diamond$$

Exercise 4.2.6

Consider the stationary Boolean model with intensity λ and random radius $0 \leq R < \infty$. Show that the Boolean model covers the whole space if and only if $\mathbb{E}[R^d] = \infty$. $\qquad \Diamond$

The *coverage probability* of a given compact set $C \subset \mathbb{R}^d$ by a random closed set Ξ is defined as $\mathbb{P}(C \subset \Xi)$. In general, it is difficult to give explicit expressions for this quantity; however, it is clear that it is not larger than $T_\Xi(C)$, and we have equality for singletons C. In the literature, there are asymptotic results for the coverage probability for the homogeneous Boolean model with Ξ equal to a centered ball of radius rR, where R is a positive random variable and r is a parameter. These results are precise in the limit $\lambda \to \infty$ of a highly dense PPP and $r \downarrow 0$ of very small communication radii. We present one such result, see [Jan86, Lemma 7.3].

Theorem 4.2.7 (Asymptotic Coverage Probability)
Assume that $d = 2$ and let $C \subset \mathbb{R}^2$ be a compact set whose boundary is a Lebesgue null set. Consider the Boolean model

$$\Xi_{BM} = \bigcup_{i \in I}(X_i + B_{rR}(0)),$$

where the random radius R satisfies $\mathbb{E}[R^{2+\varepsilon}] < \infty$ for some $\varepsilon > 0$. Put

$$\rho(\lambda, r) = \lambda r^2 \pi \mathbb{E}[R^2] - \log \frac{|C|}{\pi r^2 \mathbb{E}[R^2]} - 2 \log \log \frac{|C|}{\pi r^2 \mathbb{E}[R^2]} - \log \frac{\mathbb{E}[R]^2}{\mathbb{E}[R^2]}.$$

Then

$$\mathbb{P}(C \subset \Xi_{BM}) = \exp\left\{-\exp(-\rho(\lambda, r))\right\} \qquad as \ \lambda \to \infty, r \downarrow 0,$$

provided that $\phi(\lambda, r)$ tends to some limit in $[-\infty, \infty]$.

One can use this result for finding, for a given compact set C, the number of Poisson points, depending on the size of the local communication balls that are needed for covering C with the communication area with a certain given positive probability. Indeed, for a given $u \in \mathbb{R}$, couple λ and r such that $\rho(\lambda(r), r) \to u$ with

$$\lambda(r) = \frac{1}{\lambda r^2 \pi \mathbb{E}[R^2]}\left(u + \log \frac{|C|}{\pi r^2 \mathbb{E}[R^2]} + 2 \log \log \frac{|C|}{\pi r^2 \mathbb{E}[R^2]} + \log \frac{\mathbb{E}[R]^2}{\mathbb{E}[R^2]}\right).$$

Then, the coverage probability converges in the limit $r \downarrow 0$

$$\mathbb{P}(C \subset \Xi_{BM}) \to \exp\left\{-\exp(-u)\right\}.$$

Exercise 4.2.8
Consider the setting of Theorem 4.2.7 with $R \equiv 1$ and $C_n = [0, \sqrt{n}]^2$ a box of area n. Show that if $\pi r_n^2 = \log n + \log \log n + g(n)$ for some function g with $\lim_{n \uparrow \infty} g(n) = \infty$, then $\lim_{n \uparrow \infty} \mathbb{P}(C_n \subset \Xi_{BM}) = 1$. *Hint:* Using the Cauchy–Schwarz inequality we have

$$\mathbb{P}(|C_n \setminus \Xi_{BM}| = 0) \leq \frac{\mathrm{Var}[|C_n \setminus \Xi_{BM}|]}{\mathbb{E}[|C_n \setminus \Xi_{BM}|^2]}.$$

Use Remark 4.2.4, Exercise 4.2.5 and further bounds to obtain the result.   ◇

4.3 Long-Range Connectivity in Homogeneous Boolean Models

In this section, we consider the question of *connectivity over long distances* in the Boolean model, i.e, the question how far a message can travel through the system if it is allowed to make an unbounded number of hops. That is, we assume that a message can hop from user to user arbitrarily often, as long as it does not leave the communication area, and we ask how long the distance is that it can travel. In other words, we consider a multi-hop functionality and use the system of users as a wireless, adhoc system, which carries the message trajectories without usage of base stations. We will consider this question only for a very special, but fundamental, version of this model: the homogeneous Boolean model on the entire space $\mathbb{R}^d$ with deterministic local communication zones that are simply balls of a fixed radius. Hence, the Boolean model has only one effective parameter left, the intensity parameter of the points, but it will turn out that it gives rise to a beautiful mathematical theory that is called *continuum percolation*. We will encounter an interesting phase transition in this parameter: for large values, there is a possibility that the message can travel unboundedly far, and for small values its trajectory will always be bounded. Furthermore, we will be able to attack a number of other important quantities in later sections.

We assume that $\Phi = (X_i)_{i \in I}$ is a homogeneous PPP with intensity $\lambda \in (0, \infty)$, and the random closed set Ξ_i that we put around each user location X_i is just a deterministic closed ball $B_{R/2}(X_i)$ with a fixed radius $R \in (0, \infty)$. A message can now hop from X_i to X_j if and only if $|X_i - X_j| \leq R$, i.e., if and only if the straight line between them entirely lies in the communication area $\bigcup_{i \in I} B_{R/2}(X_i)$. This is the case if and only if the closed balls $B_{R/2}(X_i)$ and $B_{R/2}(X_j)$ intersect. Hence, we consider the Boolean model

$$\Xi_{\mathrm{BM}} = \bigcup_{i \in I} B_{R/2}(X_i) \tag{4.3.1}$$

and consider connectivity in the usual topological sense for subsets of $\mathbb{R}^d$. This notion of connectivity is suitable for transmission rules that consider message hops from any site in the area Ξ_{BM} to some device X_i, then a sequence of hops within Φ and a final hop from Φ to some site in Ξ_{BM}. Alternatively, one could also consider just the transmission through the device set Φ and connectivity within the random geometric graph Φ. This graph is an important one that we will encounter many times:

Definition 4.3.1 (Gilbert Graph)

Let $R \in (0, \infty)$, then the *Gilbert graph* with radius R on a configuration $\phi \in \mathbb{S}(\mathbb{R}^d)$ is the graph that has ϕ as its vertex set and every pair $\{x_i, x_j\}$ satisfying $0 < |x_i - x_j| < R$ as its edge. If the vertex set is given by a homogeneous PPP, we speak of a *Poisson–Gilbert graph*.

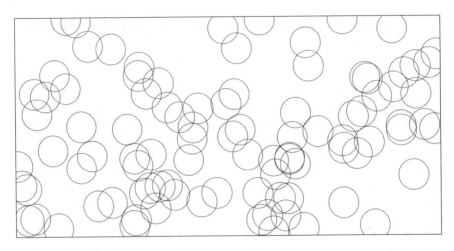

◻ Fig. 4.1 Realization of a homogeneous Boolean model

The degree $\deg(x)$ of a vertex x in a graph is defined as the number of edges that contain x. In the Gilbert graph ϕ with radius R, $\deg(x)$ is just equal to $\phi(B_R(x)) - 1$, where $B_R(x)$ is equal to the open ball with radius R centered at x.

We now have two mathematical models that express connectivity: either the Gilbert graph Φ or the Boolean model Ξ_{BM}. We will proceed in the latter model. Recall that the process is homogeneously distributed over $\mathbb{R}^d$, and that we have just two parameters, the intensity λ of Φ and the diameter R of the balls. We will write $\mathbb{P}_\lambda$ and $\mathbb{E}_\lambda$ for the probability and expectation in this model. In ◻ Fig. 4.1 we present a realization of such a Boolean model.

Let us denote by $\mathcal{C}_R(x)$ the cluster ($=$ maximal connected component) of $\Xi_{\mathrm{BM}} \cup B_{R/2}(x)$ that contains x. Then, $\mathcal{C}_R(X_i)$ is the set of those space points in $\mathbb{R}^d$ that can be reached by a multi-hop trajectory starting at X_i through Φ. One of the most decisive properties of Ξ_{BM} is whether or not it has unboundedly large components. We say that Ξ_{BM} *percolates* or that *percolation occurs* if the answer is yes. In this case, we also say that the points in the unbounded component are connected to ∞. The notion of percolation is the base of everything that follows. It gave the theory its name 'continuum percolation', since it is about the continuous space $\mathbb{R}^d$. Interestingly, the first paper that introduced this model in 1961, see [Gil61], explicitly took wireless multihop communication as the prime example and motivation. There is a theory of discrete percolation (usually motivated by water leakage through porous stones), which we will encounter in ▶ Sect. 4.4.

Remark 4.3.2 (Percolation and Message Transmission) In the event that percolation occurs, there is at least one component $\mathcal{C}$ that is unboundedly large and contains infinitely many users X_i. As a consequence, messages that are emitted from such a user can reach, at least theoretically, infinitely many other users and can travel infinitely far. Note that this does not say anything about the reach of messages that are emitted from a bounded component, which certainly also exist. In this event, the quality of transmission service is drastically better than in the complement event, such that we are highly interested in those values of the parameters that make this possible. ◇

The most important quantity is the *percolation probability*

$$\theta(\lambda, R) = \mathbb{P}_\lambda(|\mathcal{C}_R(o)| = \infty). \tag{4.3.2}$$

Since the PPP is locally finite, this is equal to the probability that $\mathcal{C}_R(o)$ contains infinitely many points of Φ.

Remark 4.3.3 (Percolation Probability via Palm Calculus) Recall the Palm distribution P_λ^o of Φ from ▶ Sect. 2.5. Then, by the stationary Mecke–Slivnyak Theorem 2.5.7, the percolation probability can be identified as

$$\theta(\lambda, R) = P_\lambda^o(|\mathcal{C}_R(o)| = \infty).$$

By the interpretation of the Palm distribution, this quantity can be also seen as the probability that the cluster $\mathcal{C}_R(X)$ has infinite Lebesgue measure, for a randomly chosen user $X \in \Phi$. The advantage of the Palm interpretation is that it can be easily extended to the setting of CPPs, see ▶ Sect. 4.8. ◇

There is actually a second natural definition of the percolation probability, which does not consider clusters of $\Xi_{\mathrm{BM}} \cup B_{R/2}(0)$, but clusters of Ξ_{BM}. Many authors consider only the latter one, in contrast to us. There are several differences and several similarities between these two theories, but at least one general inequality can be stated.

Exercise 4.3.4
Show that $\theta(\lambda, R) \geq \mathbb{P}_\lambda(|\mathcal{Z}_R(o)| = \infty) = \theta'(\lambda, R)$ where $\mathcal{Z}_R(x)$ denotes the cluster of Ξ_{BM} that contains x. ◇

Exercise 4.3.5
Show that with probability one, there are infinitely many points X_i in the Boolean model (4.3.1) with $\mathcal{C}_R(X_i) = B_{R/2}(X_i)$. ◇

Exercise 4.3.6
Show that the Boolean model based on a homogeneous PPP Φ has unbounded degree, i.e., $\mathbb{P}(\deg(X_i) < K \text{ for all } X_i \in \Phi) = 0$ for all $K > 0$. ◇

Because of the homogeneity of the PPP in space (more precisely, since the intensity measure is invariant under scaling), the influence of the parameter R can be extracted as follows:

$$\theta(\lambda, R) = \theta(\lambda R^d), \qquad \lambda, R \in (0, \infty), \tag{4.3.3}$$

where we write $\theta(\lambda) = \theta(\lambda, 1)$. Therefore, we will consider only $\theta(\lambda)$ in the following and will drop R from the notation. In view of Remark 2.4.5, it is clear that θ is an increasing function. Hence, we can define the *critical percolation threshold*

$$\lambda_{cr} = \inf\{\lambda \in (0, \infty) : \theta(\lambda) > 0\} = \sup\{\lambda \in (0, \infty) : \theta(\lambda) = 0\} \in [0, \infty], \quad (4.3.4)$$

where we put $\sup \emptyset = 0$ and $\inf \emptyset = \infty$.

Exercise 4.3.7 (Scaling of Critical Scales)

Denote by $\lambda_{cr}(R)$ the critical percolation threshold for the Boolean model with interaction radius R. Prove the scaling relation (4.3.3) and the associated scaling relation $\lambda_{cr}(R) = \lambda_{cr}R^{-d}$. Do the same scaling relations hold for θ' in Exercise 4.3.4 and its associated critical intensity?   ◊

Exercise 4.3.8

Show that $\lambda_{cr} = \infty$ for the one-dimensional Boolean model (4.3.1).   ◊

The start of the theory is the following.

> ### Theorem 4.3.9 (The Critical Threshold Is Positive and Finite)
> *For any $d \in \mathbb{N} \setminus \{1\}$, $\lambda_{cr} \in (0, \infty)$.*

The theorem says that, for some (sufficiently small) $\lambda \in (0, \infty)$, the percolation probability is zero, while for sufficiently large ones, it is positive, giving rise to non-trivial *sub- and supercritical regimes*. This is a clear manifestation of the occurrence of a *phase transition*. We give more comments in ▶ Sect. 4.5. Before we prove Theorem 4.3.9, we must have a closer look at the discrete version of percolation theory, as this will provide an inevitable mathematical base.

4.4   Intermezzo: Phase Transition in Discrete Percolation

As it is very often the case in the theory of random point processes in continuous space, proofs rest on a comparison with the (much simpler) setting of a fixed geometry. Here, instead of a PPP on $\mathbb{R}^d$, we will consider a Bernoulli field on the edges of $\mathbb{Z}^d$. That is, we put on every edge e between neighboring sites in $\mathbb{Z}^d$ a Bernoulli random variable $\xi_e \in \{0, 1\}$ with parameter $p \in [0, 1]$, and assume that all these random variables are i.i.d. Then $(\xi_e)_e$ is a Bernoulli field. The edge e is called *open* if $\xi_e = 1$ and *closed* otherwise. This turns $\mathbb{Z}^d$ into a random graph whose edges are the open nearest-neighbor edges, hence the notion of connectedness is obvious.

A realization of such a random graph is said to *exhibit percolation* if it possesses an infinite connected component. In this way, we are now working on the basic model of the theory of *discrete bond percolation*, or just *percolation*. This likewise is a beautiful mathematical theory, which studies analogous questions, but is further developed than continuum percolation, due to the simpler setting of a fixed, discrete geometry. The standard reference for this theory is [Gri89], but see also [BolRio06]. The main motivating application idea is a porous stone with a hyper-cubic microstructure, one water source in the middle and walls between neighboring cells that are permeable to water with a certain probability. One is interested in the question of whether or not the surface of the stone is wet somewhere.

Note that we have now just one parameter p, the probability for openness. We denote probability and expectation in this model by $\mathbb{P}_p$ and $\mathbb{E}_p$. We write $C(x)$ for the cluster (= maximal connected component) that contains $x \in \mathbb{Z}^d$. Similarly to (4.3.2), we introduce the percolation probability

$$\theta(p) = \mathbb{P}_p(\#C(o) = \infty) \tag{4.4.1}$$

and the critical percolation threshold

$$p_{cr} = \inf\{p \in [0, 1]\colon \theta(p) > 0\} \in [0, 1]. \tag{4.4.2}$$

Here is the start of the theory of percolation, a result that will later be instrumental for proving the continuous version, Theorem 4.3.9:

Theorem 4.4.1 (Non-trivial Critical Threshold for Bernoulli Bond Percolation)
For any $d \in \mathbb{N} \setminus \{1\}$, we have that $p_{cr} \in (0, 1)$.

Proof
The proof naturally comes in two parts. First, we prove existence of a subcritical phase, i.e., that $p_{cr} > 0$. For this, note that the event that the origin is connected to infinity is contained in the event that, for any $n \in \mathbb{N}$, a self-avoiding path of open edges with n steps starts at the origin. Let Ψ_n denote the set of self-avoiding n-step paths starting at the origin, then, for any $n \in \mathbb{N}$, we see that

$$\theta(p) \leq \mathbb{P}_p(\text{there exists } \eta \in \Psi_n \text{ such that } \xi_e = 1 \text{ for all } e \in \eta)$$

$$\leq \sum_{\eta \in \Psi_n} \mathbb{P}_p(\xi_e = 1 \text{ for all } e \in \eta) \leq \#\Psi_n p^n \leq (2dp)^n.$$

But, if $p < 1/(2d)$, then this quantity tends to zero as n tends to infinity and thus, $\theta(p) = 0$. Therefore, $p_{cr} \geq 1/(2d)$. (If one would estimate $\#\Psi_n$ against $2d(2d-1)^{n-1}$, then even the bound $p_{cr} \geq 1/(2d-1)$ would follow, which proves absence of a supercritical regime in one spatial dimension.)

The proof for the existence of a supercritical phase, i.e., that $p_{cr} < 1$, is more complicated. It suffices to prove that $\theta(p) > 0$ for $p \in (0, 1)$ close to 1. Note that if percolation occurs in dimension $d = 2$, then it also occurs for higher dimensions since there it is even easier to percolate. (This is an idea that is difficult to make precise for continuum percolation.) In other words, the critical threshold for percolation is a decreasing function of the dimension and it suffices to prove existence of a supercritical phase for dimension 2. This we will do now.

The strategy of the proof is an example of the famous *Peierls' argument*, which leverages the probabilistic costs of creating a blocking interface in the following sense. Consider the shifted lattice $\mathbb{Z}^2_* = \mathbb{Z}^2 + (1/2, 1/2)$ and call an edge e^* in $\mathbb{Z}^2_*$ closed if the unique edge e in $\mathbb{Z}^2$ that crosses e^* is open and *vice versa*. Now, if the origin is not connected to infinity, there must exist a finite blocking interface of open edges in $\mathbb{Z}^2_*$ that surrounds the origin and contains one point $(n + 1/2, 1/2)$ for some $n \in \mathbb{N}$. Thus, we can bound

$$1 - \theta(p) \le \sum_{n \in \mathbb{N}} \mathbb{P}_p\big(\exists \text{ open interface in } \mathbb{Z}^2_* \text{ around } (0, 0), \text{ passing } (n + \tfrac{1}{2}, \tfrac{1}{2})\big)$$

$$\le \sum_{n \in \mathbb{N}} \mathbb{P}_p\big(\exists \text{ open path in } \mathbb{Z}^2_* \text{ of length } 2n + 4, \text{ passing } (n + \tfrac{1}{2}, \tfrac{1}{2})\big)$$

$$\le \sum_{n \in \mathbb{N}} (4(1 - p))^{2n+4}.$$

The factor 4^{2n+4} is the number of paths of length $2n + 4$. Now, for p close to one, this sum is strictly smaller than 1 and thus, $\theta(p) > 0$. □

Exercise 4.4.2
Show that $p_{\mathrm{cr}} = 1$ for $d = 1$. ◇

Let us collect some of many interesting and important results on discrete percolation.

Remark 4.4.3 (Survey on Results on Discrete Percolation)
1. *Numerical value.* For Bernoulli bond percolation on $\mathbb{Z}^2$, the critical threshold is proven to be $p_{\mathrm{cr}} = 1/2$ based on the self-duality of $\mathbb{Z}^2$ and $\theta(1/2) = 0$. Except for a few examples (such as the triangular lattice), there is no formula for p_{cr}, and its numerical value can only be approached via simulations.
2. *Other criticality notions.* Critical behavior of Bernoulli percolation can also be based on related, but different, quantities other than the percolation function. One example is the value from which on the expected size of the cluster containing the origin is infinite: $p'_{\mathrm{cr}} = \inf\{p \in [0, 1]\colon \mathbb{E}_p[\#C(o)] = \infty\}$. For Bernoulli bond percolation on $\mathbb{Z}^2$, it is known that $p'_{\mathrm{cr}} = p_{\mathrm{cr}}$.
3. *Number of infinite clusters.* The random field $\mathbb{P}_p$ is invariant (in distribution) under lattice translations. Via Kolmogorov's 0–1 law, this implies that, in the supercritical regime $p > p_{\mathrm{cr}}$, almost surely a percolation cluster appears (i.e., $\#C(x) = \infty$ for some $x \in \mathbb{Z}^d$), since this event is measurable with respect to the tail-sigma-algebra. Further, it can be shown that this infinite cluster is unique almost surely.
4. *Sizes of finite clusters.* A lot of work has been dedicated to further understanding the clustering behavior in the two regimes. For example, in the subcritical regime, the

probability that the origin is connected to the complement of a centered box of side-length n is known to be small exponentially fast in n. This is one ingredient of the proof of $p'_{cr} = p_{cr}$. In a certain sense, the cardinality of any of the finite clusters (in both regimes) is known to be a random variable with exponential tails, which implies that the largest of the finite clusters in a box of radius n has about $\log n$ sites.

5. *Less independence.* Theorem 4.4.1 can be generalized with respect to independence. For example, if the probability p for a bond to be open is allowed to depend on neighboring bonds at distance $\leq k$ for some $k \in \mathbb{N}$, then it can be shown that there exist two critical thresholds $p^{(1)}_{cr} \leq p^{(2)}_{cr}$ such that below $p^{(1)}_{cr}$ there is no infinite cluster almost surely and above $p^{(2)}_{cr}$ there is an infinite cluster almost surely.

6. *Continuity of θ.* Another big topic in the field is to determine the continuity properties of $p \mapsto \theta(p)$. It can be shown that it is continuous in $[0, 1] \setminus \{p_{cr}\}$ and right-continuous at p_{cr}. Left-continuity at p_{cr} so far could only be established for $d = 2$ and $d \geq 11$. In particular for $3 \leq d \leq 5$, it is one of the big open questions for Bernoulli bond percolation.

7. *Behavior of θ near criticality.* It is widely believed that, on general graphs in place of $\mathbb{Z}^d$, the behavior of θ close to p_{cr} is governed by a power law that depends only on some local features of the underlying geometry. More precisely, in two dimensions it is expected that $\theta(p) = (p - p_{cr})^{\beta + o(1)}$ as $p \downarrow p_{cr}$, with the *critical exponent* $\beta = 5/36$. This has been shown rigorously only for site percolation on the triangular lattice. ◇

Exercise 4.4.4

Calculate the expected cluster size $S(p) = \mathbb{E}_p[\#C(o)]$ for Bernoulli bond percolation on $\mathbb{Z}$. Use this to determine power-law behavior of $S(p)$ as $p \uparrow 1$. ◇

Exercise 4.4.5

Show that $p \mapsto \theta(p)$ is non-decreasing. Moreover, show that $\theta(p)$ is non-decreasing in the dimension for any $0 \leq p \leq 1$. ◇

Exercise 4.4.6

Show that $\theta(p) < 1$ for any $p < 1$. ◇

Exercise 4.4.7

Show that $p \mapsto \theta(p)$ is right-continuous. *Hint:* consider the function $g_n(p) = \mathbb{P}_p$(there is a self-avoiding path of open edges of length n starting from o). ◇

In ▶ Sect. 4.5 it will be important for us to note that discrete Bernoulli percolation on $\mathbb{Z}^d$ or any other lattice can be also considered with openness attached to *sites* rather than bonds. Here it is the sites in the lattice that are independently declared open with probability $p \in [0, 1]$ and closed otherwise. The notion of clusters (= maximal connected components) is even more immediate than in the bond setting. The resulting model is called *Bernoulli site percolation*. One can be convinced that it is harder to have site-percolation than to have bond-percolation, hence the site-percolation threshold is not smaller than the bond-percolation threshold. Also for the site-version of the model, versions of Theorem 4.4.1 have been proved for various lattices, see [Gri89] and [BolRio06]. In particular, we will rely in ▶ Sect. 4.5 on the non-triviality of the percolation threshold for site percolation on the triangular lattice in two dimensions.

Exercise 4.4.8 (Analog of Theorem 4.4.1 for Site Percolation)
Prove non-triviality of the critical parameter for Bernoulli site percolation on the triangular lattice in two dimensions.   ◇

Indeed, site percolation on the triangular lattice also has the critical threshold $p_{cr} = 1/2$, see [BolRio06, Theorem 8, Chapter 5].

4.5   Proof of Phase Transition in Continuum Percolation

We are now going to apply the non-triviality of the percolation threshold for discrete percolation to the proof of the corresponding result in continuum, Theorem 4.3.9, which we are really after. However, it will not be Theorem 4.4.1 that we directly apply, but its version for site percolation on the triangular grid in two dimensions, see Exercise 4.4.8. Correspondingly, we will prove Theorem 4.3.9 only for $d = 2$.

The advantage of using the triangular lattice in the proof of Theorem 4.3.9 comes from the following. As in the proof of Theorem 4.4.1, the consideration of a discrete dual structure will be helpful. In this case, we use the regular decomposition of $\mathbb{R}^2$ into hexagons. Now, neighboring hexagons always share an edge and never only a vertex. This property (which is not shared by the dual structure for $\mathbb{Z}^2$ with site percolation) makes a more direct comparison possible between continuum percolation in $\mathbb{R}^2$ and percolation on the hexagonal structure. Furthermore, this comparison gives below bounds for the value of the critical threshold than, e.g., using site percolation on $\mathbb{Z}^2$.

Proof of Theorem 4.3.9 for $d = 2$
Here is what we have to do. Recall that on $\mathbb{R}^2$ there is a PPP $\Phi = (X_i)_{i \in I}$ given with parameter λ and the corresponding Boolean model Ξ_{BM}, the union of the balls with diameter one around each of the X_i. We have to show that, for sufficiently small λ, the probability that the connected component of Ξ_{BM} that contains the origin o with infinite Lebesgue measure is zero, and for sufficiently large λ it is positive. We will make comparisons to site percolation on a suitable triagonal lattice in two directions: first we show that site percolation in the model with sufficiently small spacing parameter implies percolation in Ξ_{BM}, and then we show that percolation in Ξ_{BM} implies site percolation in the model with sufficiently large spacing parameter. Then we use the fact that this process has a critical probability threshold in $(0, 1)$, according to Exercise 4.4.8. Even when it is equal to $1/2$ (see the remark right after the exercise), we will get qualitative estimations.

Hence, the proof naturally comes in two parts. For both directions, we will use a partition (up to boundaries, which are Lebesgue null sets) of $\mathbb{R}^2$ into open hexagons A_x^s of side-length $s > 0$ centered at some points $x \in \mathbb{R}^2$. By default, we assume that the origin is one of them. Note that the centers x form a triangular lattice $\mathcal{T}_s$, where the bonds are drawn across the edges of neighboring hexagons. We call the center x of one of these hexagons A_x^s *open* if it contains at least one Poisson point (i.e., at least one of the X_i's) and *closed* otherwise. Note that the probability for x to be open is given by

$$p_s = 1 - \exp(-\lambda |A_x^s|), \tag{4.5.1}$$

and note further that $|A_x^s| = 3\sqrt{3}s^2/2$. It is clear from the homogeneity and the Poisson nature of the PPP that openness of the sites in $\mathcal{T}_s$ defines a Bernoulli field of i.i.d. random variables. Hence, $\mathcal{T}_s$ is a Bernoulli site-percolation model.

Now, in order to show that $\lambda_{cr} < \infty$, note that any two points in neighboring hexagons have distance at most $\sqrt{13}s$. Recall that each ball around the Poisson points has diameter one. Hence, if we pick s so small that $\sqrt{13}s < 1$, then site-percolation on $\mathcal{T}_s$ implies percolation of the Boolean model Φ_{BM}. Hence, if λ is sufficiently large such that $p_s = 1 - \exp(-\lambda 3\sqrt{3}s^2/2)$ is larger than the critical threshold $1/2$ for site percolation on $\mathcal{T}_s$, then we have percolation of the Boolean model. Explicitly, we have

$$\lambda_{cr} \leq 26 \log 2/(3\sqrt{3}).$$

Now we show that $\lambda_{cr} > 0$. Note that, if $s > 1$, percolation of the Boolean model implies site-percolation on $\mathcal{T}_s$. Hence, if λ is sufficiently small such that $p_s = 1 - \exp(-\lambda 3\sqrt{3}s^2/2)$ is smaller than the threshold $1/2$, then there is no percolation of the Boolean model. Calculating again, we arrive at

$$\lambda_{cr} \geq 2 \log 2/(3\sqrt{3}),$$

which completes the proof. □

As for discrete percolation, we give now a list of further important results for continuum percolation, most of which are analogous and have proofs that are based on the discrete counterparts.

Remark 4.5.1 (More Results on Continuum Percolation)
1. *Numerical value.* The numerical value of the critical threshold is unknown in general. Rigorous bounds in $d = 2$ are $0.174 < \lambda_{cr} < 0.843$ [MeeRoy96, Section 3.9], the numerical value is $\lambda_{cr} \approx 0.6763475$, derived by computer simulations [QuiZif07].
2. *Uniqueness of infinite cluster.* In the supercritical regime, an infinitely large cluster appears almost surely in dimensions $d \geq 2$. Further, it can be shown that the infinite cluster is also unique almost surely. This uniqueness result can be substantially generalized to Boolean models based on general point processes, see [MeeRoy96, Theorem 7.4 – 7.7].
3. *Alternative critical values.* Various alternative definitions for the critical intensity can be considered. For example,

$$\lambda_1 = \inf\{\lambda : \mathbb{E}_\lambda[|\mathcal{C}_R(o)|] = \infty\},$$

$$\lambda_2 = \inf\{\lambda : \mathbb{P}_\lambda(\operatorname{diam}(\mathcal{C}_R(o)) = \infty) > 0\},$$

$$\lambda_3 = \inf\{\lambda \colon \mathbb{E}_\lambda[\mathrm{diam}(\mathcal{C}_R(o))] = \infty\},$$

$$\lambda_4 = \inf\{\lambda \colon \mathbb{P}_\lambda(\Phi(\mathcal{C}_R(o)) = \infty) > 0\},$$

$$\lambda_5 = \inf\{\lambda \colon \mathbb{E}_\lambda[\Phi(\mathcal{C}_R(o))] = \infty\},$$

where $\mathrm{diam}(C) = \sup\{|x - y| \colon x, y \in C\}$. It is known that in the Boolean model with random, independent and almost surely bounded radii, all these critical values coincide with λ_{cr}, see [MeeRoy96, Theorems 3.4 and 3.5]. In case of random independent and unbounded radii, there exist examples for which some identities fail to be true.

4. *Size of bounded clusters.* Again, as in the discrete case, in the subcritical regime, the probability that the origin is connected to the complement of a centered box of side-length n becomes small exponentially fast in n.

5. *Complement of communication zone.* It is equally interesting to study the vacant area $\mathbb{R}^d \setminus \Xi_{\mathrm{BM}}$. It can be shown that in the Boolean model there is at most one unbounded component in the vacant area.

6. *Random radii.* The above proof can be extended to also cover the case of a Boolean model with centered balls of positive random radii, independently picked for each of the Poisson points. (This process can properly be defined with the help of an independent marking of the Poisson points, see ▶ Sect. 2.4.) Let R be a random variable having the distribution of the radius of one of the balls, and denote the critical threshold by $\lambda_{\mathrm{cr}}(R)$. The first observation is that $\lambda_{\mathrm{cr}}(R)$ is not random, which can be easily shown with the help of Kolmogorov's 0–1 law. For $d \geq 2$ a supercritical regime exists, i.e., $\lambda_{\mathrm{cr}}(R) < \infty$. On the other hand, if $\mathbb{E}[R^{2d-1}] < \infty$, then $\lambda_{\mathrm{cr}}(R) > 0$. For $d = 1$, if $\mathbb{E}[R] < \infty$, then $\lambda_{\mathrm{cr}}(R) = \infty$, and if $\mathbb{E}[R] = \infty$, then $\lambda_{\mathrm{cr}}(R) = 0$. Furthermore, the map $R \mapsto \lambda_{\mathrm{cr}}(R)$ enjoys the following continuity property: If a sequence $(R_n)_{n \in \mathbb{N}}$ of random radii are uniformly bounded almost surely and converges weakly to some random radius R, then $\lambda_{\mathrm{cr}}(R_n) \to \lambda_{\mathrm{cr}}(R)$ as $n \uparrow \infty$.   ◇

Exercise 4.5.2
Prove the first part of Part 6 of Remark 4.5.1, i.e., every assertion with the exception of the continuity assertion.   ◇

4.6 Asymptotic Connectivity and Percolation

In this section, we derive a number of asymptotic results for some quantities that are important for describing large telecommunication systems:
- the percolation probability for high densities of points,
- the probability that two far distant sites are connected,
- the length of a shortest trajectory of such a connection.

We consider a standard Boolean model $\Xi_{\mathrm{BM}} = \bigcup_{i \in I} B_{R/2}(X_i)$, where $\Phi = (X_i)_{i \in I}$ is a PPP in $\mathbb{R}^d$ with intensity λ and $R \in (0, \infty)$ is the fixed diameter of the balls around the Poisson points. We write $\mathbb{P}$ and $\mathbb{E}$ for probability and expectation. We recall from

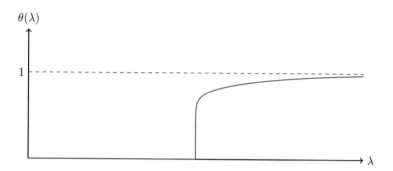

$\theta(\lambda)$

1

λ

◻ Fig. 4.2 Approximative form of the percolation probability

(4.3.2) that $\theta(\lambda, R)$ is the percolation probability, the probability that the cluster of Ξ_{BM} that contains the origin has infinitely large Lebesgue measure.

Lemma 4.6.1 (Continuity of θ and Exponential Approach to One for Large Intensities)
For any $R \in (0, \infty)$, the map $\lambda \mapsto \theta(\lambda, R)$ is continuous in $[0, \infty) \setminus \{\lambda_{\mathrm{cr}}\}$ with asymptotics

$$\lim_{\lambda \to \infty} \frac{1}{\lambda} \log(1 - \theta(\lambda, R)) = -|B_R(o)|. \tag{4.6.1}$$

The lower bound in (4.6.1) is easy to understand, since $1 - \theta(\lambda, R)$ is lower bounded by the probability that the origin is isolated, i.e., that its R-ball contains no Poisson points: $1 - \theta(\lambda, R) \geq \exp(-\lambda|B_R(o)|)$. For a full proof, see [Pen91, Corollary of Theorem 3]. Also consider [FraMee08, Theorem 2.6.3] for explanations and illustrations in two dimensions. In ◻ Fig. 4.2 we present a sketch of the graph of the percolation probability.

Exercise 4.6.2
Prove the corresponding statement for (4.6.1) in the limit of large radii and fixed intensity using scale invariance. ◇

The fact that, in the supercritical regime, with probability one, there is only one infinite component (see Remark 4.5.1) allows us to represent the probability of existence of a connection between two far distant users via the percolation probability. More precisely, let

$$p_x = \mathbb{P}(o \leftrightsquigarrow x) = \mathbb{P}(\text{there exists a path in } \Phi \cup \{o, x\} \text{ connecting } o \text{ and } x) \quad (4.6.2)$$

denote the probability that the origin is connected to a point x. We are talking here about connection between o and x in the Gilbert graph on $\Phi \cup \{o, x\}$, which has edges between pairs of points whose distance is smaller than R (see Definition 4.3.1). We write from now on $Q_s(x) = x + [-\frac{s}{2}, \frac{s}{2}]^d$ for the box centered at $x \in \mathbb{R}^d$ with side-length s.

Then we have the following result.

> **Theorem 4.6.3 (Two-Point Connectivity)**
> For any $\lambda, R \in (0, \infty)$, we have that $\lim_{|x| \to \infty} p_x = \theta(\lambda, R)^2$.

Proof

The main idea is that the only way that o and a distant site x can be connected is that each of them must belong to the infinite cluster, two events that depend only on (sufficiently large) neighborhoods of o with respect to x. The probabilities of these two events are roughly given by the percolation probability, and they become asymptotically independent if the distance is large. Let us give some details.

Let E_x and E_o denote the events that there exists a path in $\Phi \cup \{o, x\}$, starting in x and o, respectively, and leaving the boxes $Q_{|x|/3}(x)$ and $Q_{|x|/3}(0)$. Note that these two boxes are disjoint, hence E_o and E_x are independent and have the same probability. Then

$$
\begin{aligned}
|p_x - \theta(\lambda, R)^2| &\leq |p_x - \mathbb{P}(E_o \cap E_x)| + |\mathbb{P}(E_o \cap E_x) - \theta(\lambda, R)^2| \\
&\leq |p_x - \mathbb{P}(E_o \cap E_x)| + 2|\mathbb{P}(E_o) - \theta(\lambda, R)| \\
&\leq |p_x - \mathbb{P}(E_o \cap E_x)| + 2(\mathbb{P}(E_o) - \theta(\lambda, R)).
\end{aligned}
\tag{4.6.3}
$$

The second term on the right-hand side is two times the probability of the origin being connected to $\mathbb{R}^d \setminus Q_{|x|/3}(o)$ but not to infinity. This is not larger than two times the probability of the existence of a sufficiently thick interface of vacant space surrounding the origin at distance at least $|x|/3$. In the supercritical regime, the probability for this event tends to zero (even exponentially fast) as $|x| \to \infty$, see [MeeRoy96, Theorem 1].

The first summand on the right-hand side of (4.6.3), since $\{o \leftrightsquigarrow x\} \subset E_o \cap E_x$, is equal to the probability that $E_o \cap E_x$ occurs but not $\{o \leftrightsquigarrow x\}$. Note that this event is contained in the event that there exist two disjoint components in $(\Phi \cup \{o\}) \cap Q_{|x|}(o)$ of diameter at least $|x|/6$. But, the probability for this event also tends to zero (even exponentially fast) as $|x| \to \infty$, see [MeeRoy96, Lemma 4.1].

In the subcritical regime, $\theta(\lambda, R) = 0$ and $p_x \leq \mathbb{P}(E_o)$, and $\mathbb{P}(E_o)$ tend to zero as $|x| \to \infty$ (even exponentially fast), see part 4 in Remark 4.5.1.   □

Exercise 4.6.4 (Exponentially-Fast Convergence)

Prove existence of a constant $c > 0$ (depending on λ and R) such that $\lim_{|x| \to \infty} |x|^{-1} \log|p_x - \theta(\lambda, R)^2| = -c$.   ◇

An averaged version of this asymptotics, i.e., the asymptotic proportion of connected pairs of points in a large box, can be established as well. For this, let us denote by

$$\pi_s = (s^d \lambda)^{-2} \mathbb{E}\big[\#\{(X_i, X_j) \in (\Phi \cap Q_s(o))^2 \colon X_i \longleftrightarrow X_j\}\big] \qquad (4.6.4)$$

the expected number of pairs of connected Poisson points in a square of side-length s, divided by the expected number all Poisson point pairs. Then we have the following result.

> **Theorem 4.6.5 (Expected Number of Connected Pairs)**
> $\lim_{s \to \infty} \pi_s = \theta(\lambda, R)^2$.

Proof
By the Mecke–Slivnyak Theorem (2.5.7),

$$\pi_s = (s^d \lambda)^{-2} \mathbb{E}\Big[\sum_{i \in I} \sum_{j \in I} \mathbb{1}\{X_i \longleftrightarrow X_j\}\mathbb{1}\{X_i, X_j \in Q_s(o)\}\Big]$$

$$= s^{-2d}\lambda^{-1} \int_{Q_s(o)} \mathbb{E}\Big[\sum_{j \in I} \mathbb{1}\{x \longleftrightarrow X_j\}\mathbb{1}\{X_j \in Q_s(o)\}\Big] dx + (s^d \lambda)^{-1}$$

$$= s^{-2d} \int_{Q_s(o)} \int_{Q_s(o)} \mathbb{P}(x \longleftrightarrow y) \, dx dy + (s^d \lambda)^{-1}$$

$$= s^{-2d} \int_{Q_s(o)} \int_{Q_s(o)} \mathbb{P}(o \longleftrightarrow (y - x)) \, dx dy + (s^d \lambda)^{-1}$$

$$= s^{-2d} \int_{Q_s(o)} \int_{Q_s(x)} \mathbb{P}(o \longleftrightarrow z) \, dz dx + (s^d \lambda)^{-1},$$

where the additional term $(s^d \lambda)^{-1}$ comes from the fact that in Mecke–Slivnyak theorem the point x is added to the cloud of points $(X_j)_{j \in I}$.

Let a small $\varepsilon > 0$ be given. By Theorem 4.6.3, there exists an $r > 0$ such that $|\mathbb{P}(o \longleftrightarrow z) - \theta(\lambda, R)^2| < \varepsilon$ for any $z \in Q_r(o)^c$. Then we can estimate for sufficiently large s

$$|\pi_s - \theta(\lambda, R)^2| \leq s^{-2d} \int_{Q_s(o)} \int_{Q_s(x)} |\mathbb{P}(o \longleftrightarrow z) - \theta(\lambda, R)^2| \, dx dz + (s^d \lambda)^{-1}$$

$$\leq s^{-2d} \int_{Q_s(o)} |Q_s(x) \cap Q_r(o)| \, dx$$

$$+ \varepsilon s^{-2d} \int_{Q_s(o)} |Q_s(x) \cap Q_r(o)^c| \, dx + \varepsilon$$

$$\leq s^{-d}|Q_r(o)| + 2\varepsilon.$$

Since ε was arbitrary, this finishes the proof. $\qquad\qquad\qquad\square$

Exercise 4.6.6 (Polynomially-Fast Convergence)
The speed of approach of π_s towards $\theta(\lambda, R)^2$ is not exponential, but rather polynomial. Prove that $|\pi_s - \theta(\lambda, R)^2| = \mathcal{O}(s^{-2})$. $\Diamond$

We have seen that connectivity characteristics of a network represented by the Boolean model can be expressed in terms of percolation probabilities. For many types of telecommunication services, it is essential not only to know of the existence of a connection but to be able to satisfy constraints on the number of hops. Let us finish this section by explaining how a hop constraint can be captured via the *stretch factor*, a fundamental characteristic in continuum percolation.

To begin with, we extend the definition of the connectivity of two points by imposing a constraint on the number of hops. For $k \in \mathbb{N}$, we write $x \overset{k}{\longleftrightarrow} y$ if x and y can be connected in the Gilbert graph on $\Phi \cup \{x, y\}$ in at most k hops, and we write

$$p_{k,x} = \mathbb{P}(o \overset{k}{\longleftrightarrow} x). \tag{4.6.5}$$

We would like to understand the probability that any two far distant sites are connected with each other by a given number of steps. To understand the asymptotic behavior of $p_{k,x}$ for large values of k and $|x|$, we require a crucial auxiliary result known as the *shape theorem*.

Theorem 4.6.7 (Shape Theorem for the Boolean Model)
Let C denote the unique infinite connected component in the supercritical phase of percolation in the Gilbert graph on the PPP Φ with density λ and connectivity parameter $R \in (0, \infty)$. Then, for $\lambda > \lambda_{\mathrm{cr}}$ there exists a deterministic stretch factor $\rho(\lambda, R) \in (0, \infty)$ such that almost surely,

$$\lim_{\substack{|X_i - X_j| \to \infty \\ X_i, X_j \in C}} \frac{T(X_i, X_j)}{|X_i - X_j|} = \rho(\lambda, R),$$

where $T(x, y)$ denotes the smallest number of hops in the graph from x to y.

In words, the shape theorem states that the minimum number of hops to connect points in the unique infinite connected component grows linearly in the distance between the end points, and the proportionality factor depends only on the two parameters. In other words, it says that, asymptotically for large distances, the metric induced by the length of shortest paths between points (the graph distance) becomes proportional to the Euclidean metric. Sometimes $T(X_i, X_j)$ is called the *chemical distance* of X_i and X_j.

It is clear that $\rho(\lambda, R) \geq 1/R$, since one needs at least s/R steps of length R in order to travel a distance s. This extreme is reached by making all the hops in a straight line.

The proof of Theorem 4.6.7 rests on the famous *Kingman's subadditive ergodic theorem*, see [YCG11]. The definition of $T(x, y)$ generalizes to arbitrary $x, y \in \mathbb{R}^d$ via

$T(x, y) = T(q(x), q(y))$ with the point $q(x) \in C$ denoting the closest Poisson point to x that is contained in the infinite connected component. Hence, one can establish subadditivity, i.e., $T(x, y) \leq T(x, z) + T(z, y)$ for any $x, y, z \in \mathbb{R}^d$. As a consequence, $\mathbb{E}[T(o, ne_1)] \leq n\mathbb{E}[T(o, e_1)]$. Hence, in expectation,

$$\rho(\lambda, R) \leq \mathbb{E}[T(o, e_1)].$$

The numerical value of the stretch factor can only be determined via simulations. However, we have an asymptotic result on the approach of the stretch factor to one for highly dense Boolean models, established in [Hir16].

Theorem 4.6.8 (High-Density Asymptotics for the Stretch Factor)
We have that $\rho(\lambda, 1) - 1 \in \mathcal{O}(\lambda^{-1/d}(\log \lambda)^{1/d})$ as $\lambda \uparrow \infty$.

The proof rests on a discretization argument in which the scaling is recovered by considering the unlikeliness to traverse a unit box in more than one hop as the intensity tends to infinity.

Exercise 4.6.9
Establish a scaling formula for the stretch factor and use it to derive a version of Theorem 4.6.8 for growing radii and fixed intensity. ◊

With this, we can extend the representation formula in Theorem 4.6.3 to the setting of a bounded number of hops.

Theorem 4.6.10 (Asymptotic Bounded-Hop Connectivity)
Let $\lambda > \lambda_{\mathrm{cr}}$ and $R > 0$ be arbitrary, then

$$\lim_{|x| \to \infty} p_{r|x|,x} = \theta(\lambda, R)^2 \mathbb{1}\{\rho(\lambda, R) \leq r\}, \qquad r \in (0, \infty).$$

Sketch of Proof
First, the factor θ^2 appears for the same reason as in Theorem 4.6.3: (1) distant points are only connected if both are in the unique infinite connected component, and (2) the events of being in the infinite component are close to being independent if the considered points are far away from each other. By Theorem 4.6.7, the stretch factor converts constraints on the number of hops into a constraint on the Euclidean distance of the endpoints, thereby giving rise to the indicator on the right-hand side of the asserted limit. □

Another important characteristic of the Boolean model (more generally, of random and non-random graphs) is the *connective constant*, which describes the number of paths of a given length starting from the origin. We consider the Poisson–Gilbert graph formed out of the PPP $\Phi = (X_i)_{i \in I}$ with intensity λ in $\mathbb{R}^d$ with maximal bond length R (see Definition 4.3.1), and we consider the random quantity

$$W_n = \#\{\text{self-avoiding paths of length } n \text{ starting in } o\}.$$

In words, W_n is the number of n-step sequences $(o, X_{i_1}, X_{i_2}, \ldots, X_{i_n})$ such that $|X_{i_k} - X_{i_{k-1}}| \in (0, R)$ for any $k = 1, \ldots, n$ (with $X_{i_0} = o$), and all these $n+1$ sites are mutually distinct. For the Gilbert graph based on Φ replaced by the lattice $\mathbb{Z}^d$ with the usual neighborhood structure, this quantity already appeared in the proof of Theorem 4.3.9 on the nontriviality of the Bernoulli bond percolation regime on $\mathbb{Z}^d$; there it was sufficient to use some very rough bounds. The connective constant is defined as

$$\gamma = \lim_{n \uparrow \infty} \mathbb{E}[W_n]^{1/n},$$

and thus describes the asymptotic exponential growth of the expected number of self-avoiding paths beginning at the origin. In the context of message transmission in a telecommunication system, this number is relevant, for example in the analysis of (random) routing protocols, since any such path is a possible routing trajectory because loops are unwanted. Hence, the number of such paths is important for estimating the counting complexity of routing algorithms.

For any graph, the sequence $(\mathbb{E}[W_n])_{n \in \mathbb{N}}$ is submultiplicative, i.e.,

$$\mathbb{E}[W_{n+m}] \leq \mathbb{E}[W_n]\,\mathbb{E}[W_m], \qquad n, m \in \mathbb{N},$$

since every self-avoiding walk of length $n+m$ can be decomposed into two self-avoiding walks of length n and m, and the second is in distribution equal to a self-avoiding path starting from the origin. Thus, by an application of Fekete's lemma[1] , we see that γ exists and is equal to $\inf_{n \in \mathbb{N}} \mathbb{E}[W_n]^{1/n}$.

In general, it is widely open and seems unreachable to find an expression or description of the numerical value of the connective constant. For the lattice $\mathbb{Z}^d$, it is clear that each of the steps of a self-avoiding walk has at most $2d - 1$ possibilities, which gives a simple, but rough upper bound $2d - 1$ for γ. However, for the Boolean model, we are in a much better situation, as we have the following result.

> **Theorem 4.6.11** (Connective Constant for the Poisson–Gilbert Graph)
> *For any $\lambda, R \in (0, \infty)$, we have that $\gamma = \lambda|B_R(o)|$.*

The interpretation seems immediate: each site x has on an average $\lambda|B_R(o)|$ neighbors, i.e., so many possibilities to make a jump from x to another Poisson point. The reader might think that two of these possibilities have to be subducted: the current

---

[1] Fekete's lemma states that, for any subadditive sequence $(a_n)_{n \in \mathbb{N}}$ of real numbers (i.e., $a_{n+m} \leq a_n + a_m$ for any $n, m \in \mathbb{N}$), $\lim_{n \to \infty} \frac{1}{n} a_n$ exists in $[-\infty, \infty)$ and is equal to $\inf_{n \in \mathbb{N}} \mathbb{E}[W_n]^{1/n}$.

position and the Poisson point that we came from in the last jump. However, according to the Mecke–Slivnyak theorem (see Exercise 2.5.9), this is indeed not true, since the conditional distribution of $\Phi \setminus \{x, y\}$ given that x and y belong to Φ is the same as the one of Φ.

Proof

The proof rests on multiple applications of the Mecke–Slivnyak Theorem (2.5.7). We write $\sum_{i_1,\ldots,i_n \in I}^{\neq}$ for the sum on n mutually distinct indices, and we write X_{i_0} for o. More precisely, we have, for any $n \in \mathbb{N}$, that

$$
\mathbb{E}[W_n] = \mathbb{E}\left[\sum_{i_1,\ldots,i_n \in I}^{\neq} \prod_{k=1}^{n} \mathbb{1}\{X_{i_k} \in B_R(X_{i_{k-1}})\} \right]
$$

$$
= \lambda \int \mathbb{E}\left[\sum_{i_1,\ldots,i_{n-1} \in I}^{\neq} \prod_{k=1}^{n-1} \mathbb{1}\{X_{i_k} \in B_R(X_{i_{k-1}})\} \mathbb{1}\{x \in B_R(X_{i_{n-1}})\} \right] dx
$$

$$
= \lambda |B_R(o)| \, \mathbb{E}\left[\sum_{i_1,\ldots,i_{n-1} \in I}^{\neq} \prod_{k=1}^{n-1} \mathbb{1}\{X_{i_k} \in B_R(X_{i_{k-1}})\} \right]
$$

$$
= \ldots = \left(\lambda |B_R(o)| \right)^n,
$$

which gives the desired result. □

Exercise 4.6.12

Prove that for the connective constant for the Poisson–Gilbert graph, for all $\epsilon > 0$, we even have that $\limsup_{n \uparrow \infty} n^{-1} \log W_n \leq \log \gamma + \epsilon$, almost surely, using the Borel–Cantelli lemma. ◇

4.7 Bounded-Hop Percolation and Coverage

In this section, we give some first mathematical treatment and some preliminary results on coverage and connectivity properties of a communication system with many devices in a large area that is equipped with a homogeneously distributed collection of base stations. We will impose an upper bound for the number of hops of any message trajectory, and we will assume that no hop goes farther than a certain distance. Hence, the model can also be named a model of *bounded-hop percolation*, a name that was coined in the paper [Hir16]. This paper contains the first mathematical investigation of this model; it begins here.

We need to introduce notation. Let us consider a homogeneous PPP of devices $\Phi = (X_i)_{i \in I}$ with intensity $\lambda > 0$ and an independent random point process $\Psi = (Y_j)_{j \in J}$ of infrastructure, which is normalized in the sense that $1 = \mathbb{E}[\Psi(Q_1)]$. Here we recall that Q_1 is the centered unit box, and $\Psi(Q_1)$ is the number of points Y_j in Q_1. We introduce a scaling parameter $r \in (0, \infty)$ and assume that at any site rY_j of the scaled process

$\Psi_r = r\Psi$ there is a base station. The parameter r allows us to reduce the number of infrastructure nodes per unit area by taking r large. For simplicity, here we will assume that Ψ is a homogeneous PPP as well; then it has intensity one, and Ψ_r has intensity $1/r^d$.

The motivation for the model is the idea that the multi-hop functionality cannot be extended to unboundedly many hops per message for reasons of error outage, security, delivery delay, etc. Therefore, the hop number should be bounded for each message trajectory. The connections between all the base stations have an infinite capacity because of the use of wires. The drawback is that their installation and operation is enormously expensive, such that it is desirable to minimize their number. This may be achieved by putting a local multi-hop cell around each of them, which enlarges its reach. Therefore, it is important to gain information about the quality of service of each of these multihop cells, and this is addressed in the present section. One of the most obvious questions is how large the upper bound k for the hop number should be in relation to the average spacing r between the base stations. A mathematically provable answer to this question can be given in terms of asymptotics for $k, r \to \infty$ with a fixed proportion between k and r.

Recall that in (4.6.5), we have defined the probability that the origin is connected to a point in at most $k \in \mathbb{N}$ relaying hops. With the help of the shape theorem (Theorem 4.6.7), we could then derive an asymptotic formula for the two-point connection probability in Theorem 4.6.10, which essentially states that in order to have a positive connection probability the system must percolate and the number of the allowed hops should not increase too slowly as a function of the distance, compared to the stretch factor ρ. We will rely on these quantities and relations for deriving a deeper understanding of the above model.

For simplicity, we decided here to set up the model in the entire space $\mathbb{R}^d$ rather than in a compact domain, and we assumed that the two point processes of devices and base stations were independent and stationary. However, we will restrict soon to a description in a restricted, compact set. In Example 6.6.7 we will see that the compact setting is deduced from the $\mathbb{R}^d$-setting via the ergodic theorem. See also Example 6.5.5 for further averaging properties of related models.

We consider the standard Boolean model by attaching to each $X_i \in \Phi$ a ball $B_{R/2}(X_i)$ with diameter $R > 0$ representing its direct communication or coverage zone, and we do the same for all points in Ψ_r. We say that a site $x \in \mathbb{R}^d$ is *k-hop connected* to Ψ_r, and write $x \overset{k}{\leftrightsquigarrow} \Psi_r$, if x can be connected with a sequence of at most k hops of length $\leq R$ via points in X to some site in Ψ_r. Then the main object of investigation is the set

$$\Xi_k = \Xi(k, \lambda, R, r) = \{x \in \mathbb{R}^d : x \overset{k}{\leftrightsquigarrow} \Psi_r\}, \tag{4.7.1}$$

the joint coverage zone of all k-hop connected points in $\mathbb{R}^d$, or *k-hop coverage zone*. Then $\Xi_0 = \Psi_r$. Obviously, $\Xi_k \subset \Xi_{k-1}$.

Exercise 4.7.1 (Scaling Property)
Verify the following scaling formula:

$$a\,\Xi(k, \lambda, R, r) \stackrel{d}{=} \Xi(k, \lambda/a^d, aR, r/a), \qquad a \in (0, \infty),$$

where we write $\stackrel{d}{=}$ for equality in distribution. Determine precisely what invariance properties of the two point processes underlie this assertion. This means that we may put any of the three parameters λ, R and r equal to one without loss of generality. ◊

We evaluate Ξ_k in terms of the following characteristic:

$$\Theta(k) = \Theta(k, \lambda, R, r) = \lambda^{-1}\mathbb{E}\big[\#\big(\Phi \cap \Xi(k, \lambda, R, r) \cap Q_1\big)\big],$$

the expected proportion of covered points in Φ in the k-hop coverage zone in the unit cube. Let us note that $\Theta(k)$ can be also expressed in terms of the expected proportion of covered area by the k-hop coverage zone, i.e.,

$$\Theta(k) = \mathbb{E}[|Q_1 \cap \Xi(k, \lambda, R, r)|].$$

Exercise 4.7.2
Use the Mecke–Slivnyak Theorem 2.5.7 to show that

$$\Theta(k, \lambda, R, r) = \mathbb{P}(o \in \Xi(k, \lambda, R, r)) = \mathbb{E}[|Q_1 \cap \Xi(k, \lambda, R, r)|]. \qquad ◊$$

In ◘ Fig. 4.3 we give an illustration of the coverage zone extensions in which Ψ is a Cox point process (CPP, see ▶ Chap. 3 and ▶ Sect. 4.8) based on a Poisson–Voronoi tessellation (PVT, see ▶ Sect. 3.2) random environment.

We are proceeding now with deriving more interesting properties of $\Theta(k)$ that are in strong connection with coverage and percolation properties of the Boolean model. We want to give good upper and lower bounds for $\Theta(k)$. Note that the model that we are studying does not rely on connectivity of some site to infinity (with an infinite number of hops), but on connectivity of a site to a homogeneous point process with at most k hops. In the subcritical phase, we may use upper bounds that ignore the hop restriction and do not lose too much for large k, as the connectivity is bounded in some sense anyway. However, in the supercritical phase, we will get the best assertions by considering asymptotics as $k \to \infty$ and using the shape theorem for long-distance connectivity, see Theorem 4.6.7.

Recall that we denote $\lambda_{\mathrm{cr}}(R)$ the critical threshold for percolation in the Boolean model for Φ and $\mathcal{C}^{(k)}(o)$ denote the set of all points in Φ connected to the origin in less than k hops. The first result corresponds to the subcritical percolation regime and is presented in [Hir16, Theorem 1].

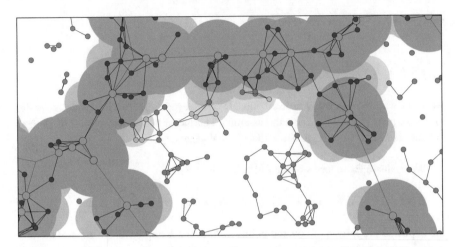

□ Fig. 4.3 Realization of Ψ (green) as a CPP based on a PVT (red) and Φ (blue and gray) in the $\mathbb{R}^2$. The gray areas indicate the k-hop coverage zones of the infrastructure nodes where $k = 1$ is dark gray, $k = 2$ is gray and $k = 3$ is light gray. Correspondingly, dark blue points are connected to Ψ directly, blue points need one intermediate hop, light blue need two hops and gray points need at least three hops

> **Theorem 4.7.3** (The k-Hop Coverage Zone in the Subcritical Regime)
> *Let $\lambda < \lambda_{\mathrm{cr}}(R)$, then*
>
> $$\Theta(k) \leq \lambda^{-1} r^{-d} \mathbb{E}[\Phi(\mathcal{C}^{(k)}(o))]. \tag{4.7.2}$$

Recall that $\mathcal{C}_R(o)$ denotes the set of points connected to the origin without hop constraint. Then, almost surely, $\Phi(\mathcal{C}^{(k)}(o)) \leq \Phi(\mathcal{C}_R(o))$ and the right-hand side of (4.7.2) is uniformly bounded by $\mathbb{E}[\Phi(\mathcal{C}_R(o))]$, which is finite in the subcritical percolation regime as mentioned in Part 3 of Remark 4.5.1.

Sketch of Proof
The proof is basically the same as the proof of [Hir16, Lemma 1] with only minor adaptations. The first main idea is clear: imagining that one point in Ψ_r is the origin, then $\Phi(\mathcal{C}^{(k)}(o))$ is the number of sites in Φ that can be reached from o by k hops. The second main idea is the mass-transport principle, which states, roughly speaking, that, due to translation invariance, counting points in $\Phi \cap Q_1$ connected to some point in Ψ_r is equivalent to counting points in Φ connected to some point in $\Psi_r \cap Q_1$. □

From Theorem 4.7.3 we can also deduce a first idea about the relation of the maximal hop number k and the intensity $1/r^d$ of the infrastructure process. Indeed, we have $\mathbb{E}[\Phi(\mathcal{C}^{(k)}(o))] \leq k^d \lambda |B_R(o)|$ and thus if $k = k(r)$ is such that $\lim_{r \uparrow \infty} k/r = c$ for some $c \geq 0$, then $\limsup_{r \uparrow \infty} \Theta(k(r), \lambda, R, r) \leq c^d |B_R(o)|$. In particular, in case that $c = 0$, if the infrastructure is reduced without allowing sufficiently many additional ad hoc hops, then the limiting expected proportion of connected devices is zero.

For the supercritical percolation regime, recall the percolation probability $\theta = \theta(\lambda, R)$ from (4.3.2) and the stretch factor $\rho = \rho(\lambda, R)$ from Theorem 4.6.7 for the Boolean model based on Φ. We write $\Psi^\theta = (Y_j)_{j \in J^\theta}$ for the thinned version of Ψ, where each $Y \in \Psi$ is kept independently with probability θ and removed otherwise. Then, we have the following result, see [Hir16, Corollary 1].

Theorem 4.7.4 (The k-Hop Coverage Zone in the Supercritical Regime)
Let $\lambda > \lambda_{cr}(R)$ and assume that $k = k(r)$ is such that $\lim_{r \uparrow \infty} k/r = c$ for some $c > 0$. Then

$$\lim_{r \uparrow \infty} \Theta(k) = \theta \, \mathbb{P}\left(o \in \bigcup_{j \in J^\theta} B_{c/\rho}(Y_j)\right). \tag{4.7.3}$$

In particular, if Ψ is even a PPP, then

$$\lim_{r \uparrow \infty} \Theta(k) = \theta\left(1 - e^{-\theta |B_{c/\rho}(o)|}\right).$$

To give some intuition for the expression on the right-hand side of (4.7.3), note that by Exercise 4.7.2 we have $\Theta(k) = \mathbb{P}(o \in \Xi(k, \lambda, R, r))$, and recall the asymptotic two point bounded-hop connection function from Theorem 4.6.10. With this in mind, note that for large r, for the origin to be connected to some point in Ψ_r, the origin needs to be in the infinite component, which gives rise to the first factor θ. On the other hand, any possible target infrastructure in Ψ_r must be part of the infinite component, which implies the fact that we have to use a θ-thinning Ψ_r^θ of the process Ψ_r. Now, the closest infrastructure in Ψ_r^θ is roughly at distance r times the closest distance of Ψ^θ to the origin, for which asymptotically at least ρr hops are required. Since asymptotically cr hops are available, a k-hop connection can only be established if $\text{dist}(o, \Psi^\theta) < c/\mu$; this gives rise to the second factor on the right-hand side of (4.7.3).

4.8 Percolation for Cox Point Processes

The proofs of the results about the percolation probability of the Boolean model in the preceding sections all benefit from the spatial homogeneity provided by the underlying PPP. However, some of them should hold also under weaker assumptions than independence. We have already introduced a way to quantify spatial correlations, at least for CPPs, via the stabilization property, see Definition 3.2.4. Let us thus present here a version of Theorem 4.3.9 for stationary stabilizing CPPs.

Before we can do this, we have to properly set up the percolation probability for stationary CPPs with intensity $\lambda \Lambda$ for some random directing measure Λ and some

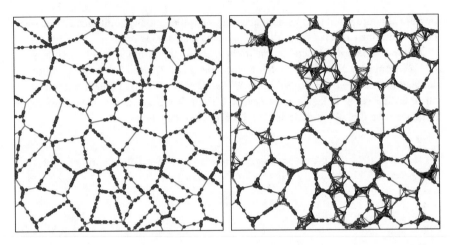

◘ Fig. 4.3 Illustration of the Boolean model for a CPP with random intensity measure based on a PVT

intensity $\lambda \in (0, \infty)$. Recall from ▶ Sect. 2.5 the definition of the Palm version of a stationary CPP Φ, i.e.,

$$E_\lambda^o[f(\cdot)] = \frac{1}{\lambda \mathbb{E}[\Lambda(Q_1(o))]} \mathbb{E}\Big[\sum_{i \in I} \mathbb{1}\{X_i \in Q_1(o)\} f(\Phi - X_i)\} \Big],$$

for all measurable $f \colon \mathcal{S}(\mathbb{R}^d) \to [0, \infty)$, where we recall the notation $Q_s(x) = x + [-s/2, s/2]^d$ for $x \in \mathbb{R}^d$ and $s > 0$.

Now we work with the Boolean model for the CPP with deterministic centered balls of radius $R \in (0, \infty)$ around each of the Poisson points. The notion of connection and clusters is the same as before, see ◘ Fig. 4.3 for an illustration.

Then the percolation probability is defined as

$$\theta(\lambda, R) = P_\lambda^o(|\mathcal{C}_R(o)| = \infty),$$

and the associated critical intensity is given by

$$\lambda_{\mathrm{cr}}(R) = \inf\{\lambda > 0 \colon \theta(\lambda, R) > 0\}.$$

Note that we cannot impose any scaling invariance that would allow us to eliminate the dependence on the reach R. We have the following result on the existence of a subcritical phase.

> **Theorem 4.8.1 (Existence of Subcritical Regime)**
> *If the random intensity measure Λ is stabilizing, then $\lambda_{\mathrm{cr}}(R) > 0$.*

The proof again rests on a comparison with Bernoulli percolation; we do not give details. This time, the comparison is first for a dependent discrete percolation system in the sense of Part 6 of Remark 4.4.3. In an additional step, the dependency can be stochastically bounded by a Bernoulli site percolation with sufficiently low openness probability.

Note that, in order to establish existence of a supercritical regime, it is not sufficient to assume stabilization for the environment. Trivially, the environment may not put positive weight on a connected region in space and consequently, even if λ is very large, the emergence of a percolation cluster is impossible. This scenario can be avoided by putting another condition on the environment which ensures enough connectivity. We write Λ_Q for the projection of the measure Λ to $Q \subset \mathbb{R}^d$, i.e., $\Lambda_Q(A) = \Lambda(Q \cap A)$.

Definition 4.8.2 (Connectedness of Random Measures)

A stabilizing random measure Λ with stabilization radii R_x, $x \in \mathbb{R}^d$, is called *asymptotically essentially connected* if for all $n \in \mathbb{N}$, whenever $\sup_{y \in Q_{2n}(o)} R_y < n/2$, we have that

1. $\text{supp}(\Lambda_{Q_n(o)})$ contains a connected component of diameter at least $n/3$, and
2. if C and C' are connected components in $\text{supp}(\Lambda_{Q_n(o)})$ of diameter at least $n/9$, then they are both contained in one of the connected components of $\text{supp}(\Lambda_{Q_{2n}(o)})$.

The PVT S of Example 3.1.4 does have this property.

Lemma 4.8.3 (The PVT Is Asymptotically Essentially Connected) *The stationary PVT on* $\mathbb{R}^d$ *is asymptotically essentially connected.*

Sketch of Proof

The proof rests again on the definition of the radius of stabilization as $R_x = \inf\{|X_i - x| : X_i \in \Phi\}$, for details see [CHJ19]. □

Exercise 4.8.4

Verify that the random environment given by random closed sets Ξ as presented in Exercise 3.1.3, where Ξ is a Boolean model, is asymptotically essentially connected if and only if the underlying Boolean model is in the supercritical percolation regime. ◊

Now we can state our result about the existence of a supercritical phase for the Boolean model based on CPPs.

Theorem 4.8.5 (Existence of Supercritical Regime)

If the random intensity measure Λ *is asymptotically essentially connected, then* $\lambda_{\text{cr}}(R) < \infty$.

Again the proof works via a discretization and comparison with a dependent Bernoulli percolation process, which can then be dominated by a supercritical independent Bernoulli percolation process, for details see [CHJ19].

Interference: Signal-to-Interference Ratios

We turn now to an important aspect of wireless communication systems, the possible disturbance of the service quality by too much noise in the system, caused by too many signals being transmitted in a vicinity. This problem, called *interference problem*, is ubiquitous in ad-hoc systems and cannot be easily neglected. In words, this phenomenon can be roughly described as follows. We imagine that from each of the locations of the devices in the system a signal is emitted. The superposition of all these signals at a given site x is called *interference* at x. If the interference is too large, then the transmission of a given signal to x can be unsuccessful.

In this section, we discuss one of the most widely used mathematical models for describing interference and show how to handle some of the most relevant properties. ▶ Section 5.1 is dedicated to the introduction of the basic tools for describing the signal strengths mathematically. In ▶ Sect. 5.2, we then present and discuss a criterion for successful transmissions in terms of the *signal-to-interference ratio (SIR)*. In ▶ Sect. 5.3 we exhibit results about the percolation properties of the variant of the Boolean model where edges are drawn based on the signal-to-interference ratio. Finally, ▶ Sect. 5.4 contains a brief discussion of protocols that are designed to help reduce interference, the *medium access control*. The standard reference on the topic of this chapter is [BacBla09a, BacBla09b], from which we learned and benefitted a lot.

As usual, in the entire chapter we consider a random point process $\Phi = (X_i)_{i \in I}$ in $\mathbb{R}^d$ and imagine that the X_i are the locations of the devices. From X_i a signal is emitted with a certain *signal power* $P_i \in (0, \infty)$. When we introduce randomness, we will mostly assume that Φ is a PPP with intensity measure μ and that $((X_i, P_i))_{i \in I}$ is a marked PPP on $\mathbb{R}^d \times (0, \infty)$.

5.1 Describing Interference

We assume that each of the network participants emits a signal. Each of the signals is propagated instantly and isotropically (i.e., radially symmetrically) into all space directions from the point of its origin. On the path of the transmission, the signal strength experiences fading effects by acoustic obstacles like the medium or by more concrete

© Springer Nature Switzerland AG 2020
B. Jahnel, W. König, *Probabilistic Methods in Telecommunications*, Compact Textbooks in Mathematics,
https://doi.org/10.1007/978-3-030-36090-0_5

ones like trees, fences, houses, and so on, which make the signal strength smaller and smaller as the distance to the transmission site increases. This fading is expressed in terms of a function $\ell: (0, \infty) \to [0, \infty)$, the *path-loss function*, i.e., $\ell(r)$ is the relative strength of the signal in a distance r from the transmitter. This function should be decreasing with $\lim_{r \to \infty} \ell(r) = 0$ and $\lim_{r \downarrow 0} \ell(r) = 1$, since the received strength cannot be larger than the emitted strength.

Very important is the kind of decay of $\ell(r)$ for large r. The general ansatz is that it should decay like a power of r, say $\ell(r) \approx r^{-\alpha}$ as $r \to \infty$, for some $\alpha > 0$. The parameter α models the average path-loss in the medium that we consider. The relation is that the more acoustic obstacles there are (e.g., in areas with a high density of houses or walls), the larger α should be picked. Often, one just determines that $\alpha > d$ to ensure integrability, see Example 5.1.1 below. Therefore, typical choices of ℓ are

$$\ell(r) = r^{-\alpha} \quad \text{or} \quad \ell(r) = (1+r)^{-\alpha} \quad \text{or} \quad \ell(r) = \min\{1, r^{-\alpha}\}. \quad (5.1.1)$$

Note that the first choice, $\ell(r) = r^{-\alpha}$, is for our application not sensible, since it explodes for small distances. On the other side, it has nice mathematical properties, for example a perfect-scaling property, such that some quantities of interest can explicitly be calculated.

Let us assume that at some site $y \in \mathbb{R}^d$ one receiver is located. The total amount of signals that it receives is equal to

$$I_\Phi(y) = \sum_{i \in I} P_i \ell(|X_i - y|), \quad y \in \mathbb{R}^d, \quad (5.1.2)$$

which is called the *interference* at y.

Let us for a while assume the simple case that $P_i = 1$ for any $i \in I$. Note that the interference at the origin, $I_\Phi = I_\Phi(o)$, is equal to $\Phi(\ell \circ |\cdot|) = \sum_{i \in I} \ell(|X_i|)$, as defined in (2.1.2). Let us have a look at its expectation.

Example 5.1.1 (Mean Interference at the Origin)

We stick to the case where $P_i = 1$. How large is the expected sum of the signals that one receiver at the origin experiences? Campbell's Theorem 2.3.1 gives

$$\mathbb{E}[I_\Phi] = \mathbb{E}[\Phi(\ell \circ |\cdot|)] = \int_{\mathbb{R}^d} \ell(|y|) \, \mu(\mathrm{d}y),$$

where μ is the intensity measure of the random point process Φ. Let us study the important special case of a standard PPP Φ with intensity λ, and let us pick one of the two bounded choices of ℓ in (5.1.1). Then, the finiteness of the integral exclusively hinges at the integrability of the integral of $\ell(|\cdot|)$ at infinity. The asymptotic for $y \to \infty$ are $\sim |y|^{-\alpha}$, hence we see that

$$\mathbb{E}[I_\Phi] < \infty \quad \Longleftrightarrow \quad \alpha > d.$$

This belongs to the reasons that often we will pick $\alpha > d$ or make the assumption that $y \mapsto \ell(|y|)$ is integrable with respect to μ, in other words that the interference is an integrable random variable. For both above choices of ℓ, it is possible without difficulties to calculate the expectation of the interference. Similarly, one can see that the perfect-scaling choice $\ell(r) = r^{-\alpha}$ leads to an integral that diverges for any value of α – either at ∞ or at zero. ◇

Exercise 5.1.2 ((Co)variance of the Interference)
Keep the assumption that $P_i = 1$. Verify that the variance of I_Φ for a homogeneous PPP Φ with intensity $\lambda > 0$ is given by $\lambda \int_{\mathbb{R}^d} \ell^2(|y|) \, dy$. Observe that this is finite if and only if $\alpha > 2d$. Furthermore, calculate in the same setting the covariance $\mathbb{E}[I_\Phi(o) I_\Phi(x)]$ and show that it tends to zero as $|x|$ tends to infinity. ◇

Example 5.1.3 (Distribution of the Interference)
Let us describe a standard procedure for describing the distribution of the interference at zero for a PPP Φ with general intensity measure μ and general path-loss function ℓ. The two basic ingredients of this procedure are (1) an approximation with the interference coming from a large ball, and (2) a decomposition according to the number of devices in that ball. However, we will see that an explicit identification of the distribution is possible only in very particular cases, even though we are still keeping the assumption that $P_i = 1$ for any $i \in I$.

Let $I_\Phi[B] = \sum_{i \in I} \mathbb{1}\{X_i \in B\} \ell(|X_i|)$ denote the interference at zero coming from all devices in a set B, and recall that $\Phi(B)$, the number of devices in B, is Poisson distributed with parameter $\mu(B)$. Furthermore, recall from Lemma 2.2.6 that, given the event $\{\Phi(B) = k\}$, the k devices in B are at independent and identically distributed sites with distribution $\mu(dx)/\mu(B)$, restricted to B. Recall that we denote by B_a the centered ball around the origin with radius a. Let us try to calculate the distribution function of I_Φ. For any $t \in [0, \infty)$, we have

$$\mathbb{P}(I_\Phi \leq t) = \lim_{a \to \infty} \mathbb{P}(I_\Phi[B_a] \leq t)$$

$$= \lim_{a \to \infty} \sum_{k \in \mathbb{N}_0} e^{-\mu(B_a)} \frac{\mu(B_a)^k}{k!} \mathbb{P}(I_\Phi[B_a] \leq t \mid \Phi(B_a) = k)$$
(5.1.3)

$$= \lim_{a \to \infty} e^{-\mu(B_a)} \sum_{k \in \mathbb{N}_0} \frac{\mu(B_a)^k}{k!} \mathbb{P}_a\left(\sum_{j=1}^{k} Y_j \leq t\right),$$

where $Y_1, \ldots, Y_k$ are k i.i.d. $(0, \infty)$-valued random variables with the distribution of $\ell(|Y|)$, where Y is a $\mu/\mu(B_a)$-distributed site in B_a. That is,

$$\mathbb{P}_a(Y \leq t) = \frac{1}{\mu(B_a)} \int_{B_a} \mathbb{1}\{\ell(|y|) \leq t\} \mu(dy), \qquad t \in (0, \infty).$$

It is not straightforward to evaluate the right-hand side of (5.1.3), but such problems are ubiquitous. The last term is the distribution function of a sum of k i.i.d. random variables and hence it looks promising to proceed instead with either the Fourier transform (characteristic function) or with the Laplace transform, since in both cases the probability term is decomposed in a product of k identical things, i.e., in a k-th power. In both cases, the k-sum can therefore be evaluated with the help of the exponential series. Let us demonstrate this for the Laplace transform. For any $s > 0$, we have

$$\mathbb{E}[e^{-sI_\Phi}] = \lim_{a\to\infty} e^{-\mu(B_a)} \sum_{k\in\mathbb{N}_0} \frac{\mu(B_a)^k}{k!} \mathbb{E}_a[e^{-s\sum_{j=1}^k Y_j}]$$

$$= \lim_{a\to\infty} e^{-\mu(B_a)} \sum_{k\in\mathbb{N}_0} \frac{\mu(B_a)^k}{k!} \mathbb{E}_a[e^{-sY}]^k$$

$$= \lim_{a\to\infty} e^{-\mu(B_a)} \exp\left(\mu(B_a)\mathbb{E}_a[e^{-sY}]\right)$$

$$= \lim_{a\to\infty} \exp\left(\int_{B_a} e^{-s\ell(|y|)}\mu(dy) - \mu(B_a)\right)$$

$$= \exp\left(\int_{\mathbb{R}^d} \left[e^{-s\ell(|y|)} - 1\right]\mu(dy)\right).$$

(This calculation is an alternate proof of Campbell's Theorem 2.3.1 for Laplace transforms.)

A similar calculation applies to Fourier transforms. Indeed, for any $\omega \in \mathbb{R}$, we see that

$$\mathcal{F}_{I_\Phi}(\omega) = \mathbb{E}[e^{i\omega I_\Phi}] = \exp\left(\int_{\mathbb{R}^d} \left[e^{i\omega\ell(|y|)} - 1\right]\mu(dy)\right).$$

Now, in order to identify the distribution of I_Φ, one has to find formulas for the inversion of the respective transforms.

An explicit evaluation of this integral seems possible only for very particular choices, for example μ the Lebesgue measure and $\ell(r) = r^{-\alpha}$. It is not clear how much one can learn form such model calculations, as the interference drastically depends on the choice of the path-loss function and on the devices that are located close to the origin. They may also produce effects that are typically unwanted and unrealistic.   ◊

Example 5.1.4 (Interference with Individual Transmission Powers)
Let us consider the case where each device X_i emits a signal with an individual transmission strength $P_i \in (0, \infty)$. We assume that, given I, the PPP $\Phi = (X_i)_{i\in I}$ and the collection $P = (P_i)_{i\in I}$ of strengths are independent, and that $(P_i)_{i\in I}$ is an i.i.d. collection. Then $((X_i, P_i))_{i\in I}$ is an independent and identically marked PPP. The interference at zero is now the random variable $I_{(\Phi,P)} = \sum_{i\in I} P_i\ell(|X_i|)$. Its Laplace transform is calculated as

$$\mathbb{E}[e^{-sI_{(\Phi,P)}}] = \exp\left(\int_{\mathbb{R}^d}\int_{(0,\infty)} \left[e^{-sp\ell(|y|)} - 1\right] K(dp)\mu(dy)\right),$$

where K is the distribution of a transmission strength. In the special case $\mu(dx) = \lambda dx$ and $\ell(r) = r^{-\alpha}$, one can calculate in an elementary way that

$$\mathbb{E}[e^{-sI_{(\Phi,P)}}] = \exp\left(\lambda c_d \int_{(0,\infty)} (ps)^{d/\alpha} K(dp)\Gamma(1 - \tfrac{d}{\alpha})\right),$$

where Γ denotes the Gamma function and c_d is the volume of the d-dimensional unit ball, for example $c_2 = 2\pi$. ◊

5.2 Signal-to-Interference Ratios

The interference $I_\Phi(X_k)$ at the location of the device X_k is the superposition of all the signals at this site from all the others. The device at X_k is interested to successfully understand a signal that comes from a particular device X_i. The interference makes it hard to be successful. A precise mathematical criterion for this success is given in terms of the *signal-to-interference-and-noise ratio* (SINR),

$$\mathrm{SINR}(X_i, X_k, \Phi) = \frac{P_i \ell(|X_i - X_k|)}{N + \sum_{j \in I \setminus \{i,k\}} P_j \ell(|X_j - X_k|)}, \qquad i \neq k \in I, \tag{5.2.1}$$

where $N \in (0, \infty)$ is the general *noise* in the system, $P_j \in (0, \infty)$ is the signal power of the j-th device, and ℓ is a path-loss function as in ▶ Sect. 5.1. In words, $\mathrm{SINR}(X_i, X_k, \Phi)$ is the quotient of the (wanted) signal strength that is received at X_k from the device X_i and the (unwanted) total sum of the basic noise and all the other signal strengths transmitted from all the other devices. Hence, the sum in the denominator is not really equal to the interference at X_k, but the difference is only two summands, since we do not include the signals emitted from X_k nor X_i.

In the literature, it appears to be a matter of taste whether or not to exclude the summands for $j = k$ and $j = i$. For this, note that, formally, the summand for $j = k$ can be seen as part of the noise term N and in terms of modeling, usually it is safe to assume that the device X_k can efficiently filter its own signal, such that it does not contribute to the interference. For the summand $j = i$, we will show in Remark 5.3.1 that it can also be added without changing the model once another system parameter is properly adjusted.

If we want to neglect the general noise, then we put $N = 0$ and call the quantity in (5.2.1) the *signal-to-interference ratio* (SIR) and write it as $\mathrm{SIR}(X_i, X_k, \Phi)$. Certainly, we can further simplify by putting $P_i = 1$ for any i.

For a receiver at a site $y \in \mathbb{R}^d$, one can define a variant $\mathrm{SINR}(X_i, y, \Phi)$ of the SINR by replacing X_k by y and extending the sum in the denominator over all $j \in I \setminus \{i\}$ if $y \neq X_k$. The success criterion is then formulated as:

$$\text{a receiver at } y \text{ can detect the signal from } X_i \iff \mathrm{SINR}(X_i, y, \Phi) \geq \tau, \tag{5.2.2}$$

where $\tau \in (0, \infty)$ is a technical constant, which measures the fineness of the ability to filter the wanted signal from the unwanted ones. The area of sites that can detect the signal emitted from a given Poisson site $X_i \in \Phi$,

$$\mathcal{C}_{X_i} = \{y \in \mathbb{R}^d : \mathrm{SINR}(X_i, y, \Phi) \geq \tau\} \tag{5.2.3}$$

is called its *SINR cell*; we give an illustration in ◘ Fig. 5.1.

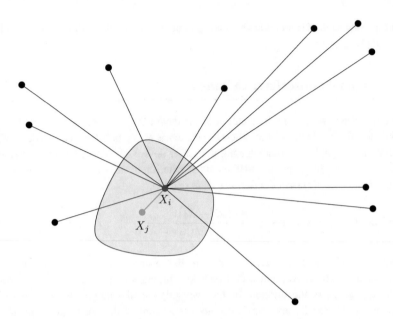

⧠ Fig. 5.1 Illustration of the SINR cell (light green) of a device X_i (blue). The device X_j (green) can transmit to X_i while other devices (black) create interference at X_i (black lines), but cannot successfully transmit

Remark 5.2.1 (SINR Cells Converge Towards Voronoi Cells) How can one roughly get an idea about the shape of an SINR cell around a given X_i? One is tempted to say that, roughly, it consists of the set of those sites that are closer to X_i than any of the other Poisson points, since one would think that they have the best chances to be connected to X_i. In other words, one could think that the SINR cell around X_i is similar to the Poisson–Voronoi cell around X_i $z(o) = \{y \in \mathbb{R}^d : |y| \le \inf_{j \in I} |y - X_j|\}$ defined in (3.2.1). This intuition can be made rigorous in one particular limiting setting. Indeed, we consider the path-loss function $\ell(r) = (1 + r)^{-\alpha}$ and positive signal powers, and we consider the limit as $\alpha \uparrow \infty$. Let us denote the SINR cell at zero by $\mathcal{C}_o^\alpha$ and the interference by I_Φ^α, to make the dependence on α explicit. We want to make understandable that indeed

$$\lim_{\alpha \uparrow \infty} \mathcal{C}_o^\alpha = z(o),$$

in the sense of convergence of random closed sets. For a full proof see [BacBla09a, Proposition 5.5.11].

Convergence of closed sets, or Painlevé–Kuratowski convergence of closed sets, see [BacBla09a, Definition 5.5.1], is defined as follows. A sequence $(A_n)_{n \in \mathbb{N}}$ of closed sets in $\mathbb{R}^d$ converges to the closed set $A \subset \mathbb{R}^d$ if (1) for all $x \in A$ there exists a sequence $(x_n)_{n \in \mathbb{N}}$ with $x_n \in A_n$ for all but finitely many n and $x_n \to x$ and (2) for any subsequence $(A_{n_k})_{k \in \mathbb{N}}$ and any sequence of points $(x_k)_{k \in \mathbb{N}}$ with $x_k \in A_{n_k}$ and $x_k \to x$ we have $x \in A$.

Now, to see the convergence of the SINR cell towards the Voronoi cell intuitively, note that (by some little exercise) it is seen that

$$\lim_{\alpha\uparrow\infty} I_\Phi^\alpha(y)^{1/\alpha} = \frac{1}{1 + \inf_{i\in I}|y - X_i|},$$

and thus the interference at y is dominated by the closest Poisson point to y. Neglecting the noise term, we can hence write for any $y \in \mathbb{R}^d$

$$\lim_{\alpha\uparrow\infty} \mathbb{1}\{\mathrm{SINR}(X_i, y, \Phi) \geq \tau\} = \lim_{\alpha\uparrow\infty} \mathbb{1}\left\{\frac{P_i^{1/\alpha}(1+|y|)^{-1}}{I_\Phi^\alpha(y)^{1/\alpha}} \geq \tau^{1/\alpha}\right\}$$

$$= \mathbb{1}\left\{\frac{(1+|y|)^{-1}}{(1 + \inf_{i\in I}|y - X_i|)^{-1}} \geq 1\right\}$$

$$= \mathbb{1}\{|y| \leq \inf_{i\in I}|y - X_i|\}.$$

Now it is seen that the Poisson–Voronoi cell appears in the limit. The proof now consists in checking the conditions for convergence of closed sets and proving the steps in the above calculation. ◇

Exercise 5.2.2
Formulate and prove a similar result as in Remark 5.2.1 for $\ell(r) = \exp(-\alpha r)$. ◇

The concept of SINR cells is obviously a natural extension of the Boolean model that is constructed out of the local communication areas Ξ_i in ▶ Sect. 4.1, under consideration of interference constraints. Analogously to the Boolean model in (4.3.1), the union of the SINR cells,

$$\Xi_{\mathrm{SINR}} = \bigcup_{i\in I} C_{X_i}, \tag{5.2.4}$$

is called the *SINR model*. However, note that the Boolean model is by no means a particular case of the SINR model, not even for choices like $\ell(r) = \mathbb{1}\{r \leq R\}$ for some R. The SINR does not have the independence properties of the Boolean model.

We define the *SINR coverage probability*

$$p_o(y) = P^o(y \in C_o) = T_{\Xi_{\mathrm{SINR}}}(\{y\}), \qquad y \in \mathbb{R}^d, \tag{5.2.5}$$

as the probability (under the Palm measure) that a given site y can be reached by a signal emitted from the origin o; recall also the capacity functional introduced in (4.2.2).

Remark 5.2.3 (Calculating the Coverage Probability) Let us elaborate a bit on explicit calculations for the SINR coverage probability,

$$p_o(y) = \mathbb{P}\left(P_o\ell(|y|) \geq \tau\left(N + \sum_{j\in I} P_j\ell(|X_j - y|)\right)\right).$$

This probability involves the PPP Φ, the random noise N and the i.i.d. signal strengths P_i, where we assume independence of all these quantities. It is difficult to obtain more explicit formulas for the SINR coverage probability in general. In [BacBla09a] formulas involving Fourier transforms of the interference and of the signal-strength variables are derived. Let us present here an approach for the case that the signal strengths P_i are exponentially distributed with parameter c. In this case we are in the lucky situation that we can employ the Laplace transforms of N and of the interference $I_{(\Phi,P)}(y) = \sum_{j \in I} P_j \ell(|X_j - y|)$ as follows. We write $\mathcal{L}_Z(t) = \mathbb{E}[\exp(-tZ)]$, $t \in [0, \infty)$, for the Laplace transform of a random variable Z.

$$
\begin{aligned}
p_o(y) &= \int_0^\infty \mathbb{P}(P \geq \tau s / \ell(|y|)) \, \mathbb{P}(N + I_{(\Phi,P)}(y) \in \mathrm{d}s) \\
&= \int_0^\infty \exp\left(- c\tau s / \ell(|y|)\right) \mathbb{P}(N + I_{(\Phi,P)}(y) \in \mathrm{d}s) \\
&= \mathcal{L}_{N + I_{(\Phi,P)}(y)}\left(c\tau / \ell(|y|)\right) \\
&= \mathcal{L}_N\left(c\tau / \ell(|y|)\right) \mathcal{L}_{I_{(\Phi,P)}(y)}\left(c\tau / \ell(|y|)\right),
\end{aligned}
$$

where we use that the Laplace transform of a sum of two independent random variables is equal to the product of the two Laplace transforms. For calculating the transform of $I_{(\Phi,P)}(y)$, one might use Example 5.1.4. In the simple case where $N = 0$ and $\ell(r) = r^{-\alpha}$ for some $\alpha \in (0, \infty)$, one can derive explicit results. $\quad\Diamond$

Exercise 5.2.4
Derive in a similar fashion as in Remark 5.2.3 an expression for the covariance $p_o(x, y) = P^o(\{x, y\} \subset C_o)$ and give conditions under which $\lim_{|x-y| \uparrow \infty} p_o(x, y) = 0$. $\quad\Diamond$

5.3   SINR Percolation

Given the random point cloud Φ, the SINR gives us a more refined and more realistic rule to build a random graph with vertices given by Φ. The classical Boolean model, where an edge is drawn between two points entirely based on their mutual distance, can now be replaced by the *SINR model*, where a directed edge is drawn from X_i to X_k in Φ if $\mathrm{SINR}(X_i, X_k, \Phi) \geq \tau$. In many systems, one considers a message transmission $X_i \to X_k$ successful only if also a confirmation message $X_k \to X_i$ is successfully transmitted. Hence, it makes good sense to introduce symmetry in the edges and consider the undirected graph having an edge between X_i and X_k in Φ if both $\mathrm{SINR}(X_i, X_k, \Phi) \geq \tau$ and $\mathrm{SINR}(X_k, X_i, \Phi) \geq \tau$. In this way, we obtain a graph structure on the random point process Φ, which is symmetric, i.e., the edges are undirected. Edges are drawn for bonds along which the direct communication is successful in both directions under consideration of interference. This is under the assumption that, at any considered time instant, each node transmits precisely one message.

Let us introduce an additional parameter $\gamma > 0$, the *interference-cancellation factor*, which allows us to tune the interference and write

$$\text{SINR}_\gamma(X_i, X_k, \Phi) = \frac{P_i \ell(|X_i - X_k|)}{N + \gamma \sum_{j \in I \setminus \{i\}} P_j \ell(|X_j - X_k|)}, \qquad y \in \mathbb{R}. \tag{5.3.1}$$

We call the corresponding graph the *SINR$_\gamma$ model*. Note that we now included the summand $j = k$ in the interference; compare to (5.2.1).

Remark 5.3.1 (Relative Versus Total Interference) The exclusion of the transmitter X_i in the interference term in the denominator in (5.3.1) is a standard model assumption. However, under a suitable change of the connectivity parameter τ, the reduced interference $\sum_{j \in I \setminus \{i\}} P_j \ell(|X_j - y|)$ can be replaced by the total interference $I_{(\Phi, P)}(y)$. Indeed, we have the equivalence

$$\text{SINR}_\gamma(X_i, y, \Phi) \geq \tau \qquad \Longleftrightarrow \qquad \frac{P_i \ell(|X_i - X_k|)}{N + \gamma I_{(\Phi, P)}(y)} \geq \frac{\tau}{1 + \tau \gamma},$$

since the left-hand condition is equivalent to

$$P_i \ell(|X_i - y|) \geq \tau \left(N + \gamma \sum_{j \in I \setminus \{i\}} P_j \ell(|X_j - y|) \right)$$

$$= \tau \left(N + \gamma I_{(\Phi, P)}(y) - \gamma P_i \ell(|X_i - y|) \right),$$

which is equivalent to $P_i \ell(|X_i - y|) \geq \frac{\tau}{1 + \tau \gamma} \left(N + \gamma I_{(\Phi, P)}(y) \right)$. ◊

In contrast to the Boolean model, the SINR$_\gamma$ model has far-reaching correlations, and one might think that its mathematical treatment should be much more difficult than the one of the Boolean model. This is in general true, but with respect to the degrees of its nodes, the SINR$_\gamma$ model has a rather simple property. Indeed, while the vertices in the Boolean model in an infinite space have unbounded degrees (in the sense that, with probability one, there exists an infinite sequence of devices $(X_{i_n})_{n \in \mathbb{N}}$ such that each X_{i_n} has at least n neighbors), see also Exercise 4.3.6, in the SINR$_\gamma$ model this cannot happen.

Lemma 5.3.2 (The SINR$_\gamma$ Model Has Bounded Degree) *Let $\gamma > 0$, then in the SINR$_\gamma$ model, with probability one, each node has at most $1 + 1/(\gamma \tau)$ neighbors.*

Proof
Let X_0 be any device in Φ and denote by N_0 the number of its neighbors in the SINR$_\gamma$ graph. If $N_0 \leq 1$, there is nothing to show. If $N_0 > 1$, then with probability one, there exists a neighbor of X_0 with smallest signal power, i.e., there exists $X_1 \in \Phi$ such that for all $i = 2, \ldots, N_0$,

$$P_1 \ell(|X_1 - X_0|) \leq P_i \ell(|X_i - X_0|).$$

But then,

$$P_1 \ell(|X_1 - X_0|) \geq \tau \left(N + \gamma \sum_{j \in I \setminus \{1\}} P_j \ell(|X_j - X_0|) \right)$$

$$\geq \tau \gamma (N_0 - 1) P_1 \ell(|X_1 - X_0|),$$

and thus $N_0 \leq 1 + 1/(\gamma \tau)$.   □

In particular, if $\gamma > 1/\tau$, then each node has at most one neighbor and the SINR_γ model has components of size at most two. In that sense, the SINR_γ model can be used as an approximation of a Boolean model, as γ tends to zero. This approximation has the property that edges of nodes with high degrees in the limiting Boolean model are only gradually added to the graph.

Exercise 5.3.3
Assume an SINR_γ model based on a stationary PPP Φ without noise and fading variables and $\ell(r) = r^{-\alpha}$ in two dimensions. Verify that the expected number of Poisson points that can transmit to the origin is given by $\tau^{-1/\alpha} \lambda 2\pi \mathbb{E}[I_\Phi^{-1/\alpha}]$. $\Diamond$

Exercise 5.3.4
Show that the random measure $\Lambda(dx) = \nu_1(S \cap dx)$, where S is the set of edges in the SINR_γ model based on a homogeneous PPP, is stabilizing for any choice of γ, when noise and fading variables are neglected. Under what conditions on noise and fading variables is Λ still stabilizing? $\Diamond$

Remark 5.3.5   (SINR$_\gamma$ Models Versus Boolean Models)
1. *No interference.* For $\gamma = 0$, the interference is neglected and the SINR_γ model becomes a Boolean model with potentially random radii depending on the fading variables and on N. Assuming that all devices transmit at some fixed maximum power $P \in (0, \infty)$, the radius of interaction is then given by

   $$R = \sup\{r \geq 0 \colon \ell(r) \geq \tau N/P\},$$

   where randomness only enters via the noise variable N. More specifically, for $\ell(r) = r^{-\alpha}$ we have that $R = (P/(\tau N))^{1/\alpha}$. Recalling statement 5 in Remark 4.5.1, under some assumptions on the distribution of N, there exists a non-trivial $\lambda_{\text{cr}}^* \in (0, \infty)$ marking the phase-transition point that separates a sub- and supercritical regime of percolation for this Boolean model.
2. *Ordering of graphs.* Note that for any $\gamma > 0$, realization-wise, the SINR_γ model is contained in the SINR_0 model since interference can only decrease the connectivity. Hence, for $\lambda < \lambda_{\text{cr}}^*$, the SINR_γ model is subcritical.
3. *Existence of critical value.* For $\lambda > \lambda_{\text{cr}}^*$ we know that
   (a) the SINR_0 model is supercritical and
   (b) for $\gamma > 1/\tau$, by Lemma 5.3.2, the SINR_γ model is subcritical.

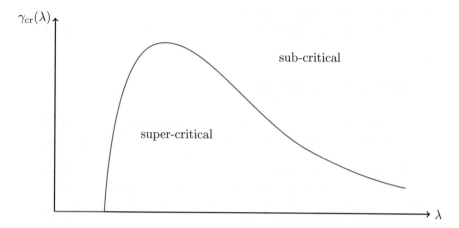

□ Fig. 5.2 Sketch of the phase diagram for percolation of the SINR$_\gamma$ model

Hence, there must exist a critical $0 \leq \gamma_{\mathrm{cr}}(\lambda) \leq 1/\tau$ at which the SINR$_\gamma$ model has a phase transition of percolation. ◇

The following theorem makes statements about $\gamma_{\mathrm{cr}}(\lambda)$ from the previous remark under rather strong model assumptions. A proof can be found in [DBT05].

Theorem 5.3.6 (SINR$_\gamma$ Percolation)
Let $d = 2$ and assume $P_i = P$ and N to be fixed positive and finite numbers. Further assume that there exist $0 < \delta < \beta$ and $M > \tau N/P$ such that the path-loss function satisfies $\ell(r) = 0$ for all $r > \beta$ and $\tau N/P < \ell(r) < M$ for all $r < \delta$. Then, there exists $\lambda' < \infty$ and a function $\gamma_{\mathrm{cr}} : [\lambda', \infty) \to (0, \infty)$ such that, for $\lambda > \lambda'$ and $\gamma < \gamma_{\mathrm{cr}}(\lambda)$ there exists almost surely an infinite component in the SINR$_\gamma$ model.

A more complete statement for the existence of a super-critical phase in the presence of interference is given in [DFMMT06], where in particular the condition of the boundedness of the support of the path-loss function is replaced by an integrability requirement together with a condition of strict monotonicity. Let us stress that, with interference, the percolation probability is not a monotone function of the intensity anymore. In fact, for a sufficiently large intensity, the connectivity of the model begins to decrease due to the effect of interference. The general picture is summarized in □ Fig. 5.2.

Remark 5.3.7 (SINR$_\gamma$ Percolation for CPPs) Let us finally mention that the SINR$_\gamma$ percolation result has been substantially generalized with respect to the underlying point process and the SINR. More precisely, in [Tob18] it has been shown that in case the PPP Φ is replaced by a stationary CPP with asymptotically essentially connected driving measure, in at least two spatial dimensions, non-trivial sub- and supercritical percolation regimes can be observed in the large intensity regime for sufficiently small but positive γ. The proofs rely on similar but stronger conditions compared to the ones used in [DFMMT06], and also new arguments are used. $\Diamond$

5.4   Medium Access Control

Let us close this chapter by hinting towards another research direction within the field of adhoc telecommunication networks, which is amenable to an analysis via methods of stochastic geometry. That is the study of protocols that manage the network participants' access to the system, i.e., the medium. Any such protocol is part of the so-called *medium access control* (MAC) of the network. We have seen in the previous sections that the activity of many participants leads to large interferences and consequently to a situation where many messages may not be transmittable due to too low SINR. However, in reality, the sum over j in the denominator of the SIR in (5.2.1) should actually extend only over the set of those devices that transmit precisely at the time that we consider, not necessarily over all participants in the system. If one considers only one time instant, one could certainly impose that Φ is meant to be the set of all the transmitting devices, but in a more complex model that takes a time development into consideration, certainly this point set can vary from time to time. In such a model, one could interpret Φ as the set of participants of the system, and at a given time instant one has to determine which ones of them are transmitting.

If many devices want to transmit messages, then it will be absolutely necessary to distribute the times of transmissions in such a way that, at each of the time instances, only a carefully chosen part of the messages makes a hop, such that the sum in the denominator of (5.2.1) is not too large and the criterion in (5.2.2) is met for many or even all those X_i that transmit. A good strategy will typically allow only a certain density of devices to transmit at a given time instance, and those who transmit should be spatially widely spread. Finding a good algorithm to decide which devices are allowed to transmit, is an important problem and various protocols can be considered featuring various levels of complexity.

We focus in this section on decentralized protocols, where the (random) decisions are made based on local information, allowing for a reasonable amount of independence in the system. This ansatz makes it especially easy for the devices to make individual decisions without the need for global knowledge about the system or to calculate highly complex optimal solutions to the interference problem and then to inform each device about the optimal decision. The simplest algorithm of that type may be that each participant independently decides to transmit with a fixed probability $p \in (0, 1)$, not depending on anything else, the *ALOHA protocol*, which was introduced in the early 1970s, see [BacBla09b] and references therein. For a PPP Φ, then the set of transmitting devices is nothing but the p-thinning of Φ in the sense of Lemma 2.2.15.

Another important protocol, where decision for the transmissions are based on some knowledge of the local environment of the devices, is called *carrier sense multiple access protocol* (CSMA), see also [BacBla09b]. Here, the probability $p \in (0, 1)$ is used to thin the transmitter process in a similar fashion as in the type 2 Matérn hard-core point process, see Example 2.4.9. Let us at least mention here another important protocol, the protocol of *code division multiple access (CDMA)*, in which receivers are equipped with a filtering algorithm that allows them to recognize the desired signal and consider the other signals as noise, see [BacBla09b] for details.

Considering the spatial ALOHA and CSMA protocols, finding the best thinning can be based on a number of performance metrics, such as the probability of coverage of a typical device in the system conditioned to be a transmitter, the expected Shannon throughput of the typical device, the mean transmission progress made by a typical device, the mean transport of a typical device, etc. Another important characteristic is the expected throughput, i.e., the amount of messages that are successfully transmitted and received, under the consideration of interference. In the simplest case of the spatial ALOHA protocol, it is easy to see that small values of p lead to a small number of transmission attempts, whereas large values lead to a high risk of disappointment for interference reasons. Already in this simple setting, it is seen that the interference has to be controlled in any spatial location and all the interferences are highly dependent random variables. This gives an indication of the enormously high complexity of this matter. The analysis of approximations for limiting settings is just one of the many mathematical tasks to be done in that direction.

Large Systems: Convergence of Point Processes

We are interested in *large* telecommunication systems, i.e., systems with many participants. In order to obtain a clear picture of such a complex situation, we need simplifying limiting formulas. The main two limiting situations that we consider are

- *high spatial density* of many devices in a compact communication area,
- *large-space averages* of many devices in a large box.

In the first setting, many devices are densely packed and form a more or less homogeneous cloud, which we will approximate with a measure on the communication area in ► Sect. 6.2. In this limiting setting, only macroscopic observables of the cloud will survive in the description, which reduces the complexity of the picture in a similar way as in the law of large numbers, nevertheless keeping a great deal of spatial details.

In the second setting, every single device and all the microscopic details can be resolved, and we will be interested in large-space averages. Hence, the appropriate setting to consider is the one of stationary point processes in the entire space $\mathbb{R}^d$ and their ergodic behavior. To set the stage for such kind of limiting scenarios, we first recall, in the simpler and more widely known situation of stationary processes with index set $\mathbb{N}$, Birkhoff's ergodic theorem and related issues in ► Sect. 6.3. We identify limiting spatial averages in ► Sect. 6.4 using the d-dimensional, spatially continuous version of the ergodic theorem, the famous Wiener ergodic theorem. The limit in Wiener's ergodic theorem is a priori random, but under the assumption of ergodicity of the underlying point process, it is deterministic. At the end of ► Sect. 6.4, we therefore give an extensive account on ergodicity and related issues like mixing properties. In ► Sect. 6.5, we prepare for applications of Wiener's ergodic theorem to telecommunication models by showing how to control of boundary terms, and we discuss a number of use cases. The important extension of Wiener's ergodic theorem to marked point processes is detailed in ► Sect. 6.6, together with a number of examples of use cases. Formulations for and extensions to stationary empirical fields and individual empirical fields are given in ► Sect. 6.7. More applications are broadly discussed in ► Sect. 6.8.

For a proper mathematical treatment for both main scenarios, we first need to discuss convergence issues for (random) point measures, which we will do in ► Sect. 6.1.

© Springer Nature Switzerland AG 2020
B. Jahnel, W. König, *Probabilistic Methods in Telecommunications*, Compact Textbooks in Mathematics,
https://doi.org/10.1007/978-3-030-36090-0_6

6.1   Convergence of Random Point Measures

In this section, we present some useful general tools for a mathematical treatment of convergence of point measures, in particular random ones. See also some standard texts on this subject like [Bil68] for general material about convergence of measures, and [DalVer03, Appendix A2], [MKM78] and [Res87], or [Kal97, Appendix A2] for particular consideration of point measures. The basis was laid in ▶ Sect. 2.1, where the distribution of a random point process was discussed. Here we proceed one step further and provide tools for characterizing convergence. First we discuss convergence of the type that we will be concerned with in ▶ Sect. 6.2, high-density limits, and then we turn to convergence notions that are suitable for ergodic limits as studied in ▶ Sect. 6.4.

As before, we fix a measurable set $D \subset \mathbb{R}^d$. First we turn to the most popular notion for convergence of point measures $\sum_{i \in I} \delta_{x_i}$ on configurations $(x_i)_{i \in I} \in \mathbb{S}(D)$ towards other measures, the vague convergence. Here it is important for us to include also the case where the limiting measure is not necessarily a point measure anymore, but still a measure that assigns to compact sets finite values, i.e., a Radon measure. The appropriate general setting is the setting where D is just some locally compact topological Hausdorff space that satisfies the second countable axiom, but, as in the preceding sections, we just keep this in mind and proceed with a measurable set $D \subset \mathbb{R}^d$.

Let $\mathfrak{R}(D)$ denote the set of all Radon measures on D, then the vague topology that we introduced in Definition 2.1.1 on $\mathbb{S}(D)$ easily extends to $\mathfrak{R}(D)$. Indeed, vague convergence is defined by convergence of all the test integrals against all the continuous bounded functions $D \to \mathbb{R}$ with compact support. More explicitly, a sequence $(\pi_n)_{n \in \mathbb{N}}$ of Radon measures on D converges vaguely towards a Radon measure π if $\lim_{n \to \infty} \langle f, \pi_n \rangle = \langle f, \pi \rangle$ for any $f \in \mathcal{C}_c(D)$, where we write $\langle f, \pi \rangle$ for the integral of f with respect to π. A variant of the *Portmanteau theorem*[1] (see [Kal97, Theorem A2.3] or [DalVer03, Proposition A2.6.II.]) says that this is the same as convergence of the measures of compact subsets whose boundary is a nullset with respect to the limiting measure, i.e., $\lim_{n \to \infty} \pi_n(A) = \pi(A)$ for any compact $A \subset D$ with $\pi(\partial A) = 0$.

Example 6.1.1   (Locality of Vague Convergence)
Vague convergence is a local notion and says nothing about the total mass, as one sees in the examples that δ_n on $D = \mathbb{R}$ converges vaguely to the zero measure as $n \to \infty$ and that the measure on $D = \mathbb{R}$ with Lebesgue density $\mathbb{1}_{[-n,n]}$ converges vaguely towards the Lebesgue measure.   ◊

Remark 6.1.2   (Metrizability)   There is a metric on $\mathfrak{R}(D)$ that induces the vague topology, see [DalVer03, A2.5]. Hence, the topological space of such measures is even a metric space.
◊

---

[1] The Portmanteau theorem states that weak convergence of probability measures (defined by convergence of all the integrals against continuous bounded functions) is equivalent to convergence of their masses of any measurable set whose boundary is a nullset with respect to the limiting measure.

Remark 6.1.3 (Vague Measurability) On the topological space $\mathfrak{R}(D)$ (equipped with the vague topology), there is of course a measurable structure that is given by the Boreal σ-field. The maps $\mu \mapsto \mu(A)$ with $A \subset D$ measurable and compact form a basis of this Boreal σ-field; i.e., it is the smallest σ-field that makes these maps measurable (see [DalVer03, Theorem A2.6.III]). ◇

Remark 6.1.4 (Relative Compactness of Sets of Measures) It can be deduced from Prohorov's theorem[2] that a sequence $(\mu_n)_{n \in \mathbb{N}}$ of Radon measures on D is relatively compact in the vague topology (i.e., each subsequence contains a further subsequence that vaguely converges) if and only if, for any relatively compact set $A \subset D$, the sequence $(\mu_n(A))_{n \in \mathbb{N}}$ is bounded. ◇

Now we turn to sequences of *random* Radon measures, i.e., sequences of random variables taking values in the set $\mathfrak{R}(D)$, and characterize limits. A very natural sense of convergence of $\mathfrak{R}(D)$-valued random variables is their *convergence in distribution* or *weak convergence*. This is the sense of convergence that we will be relying on in ▶ Sect. 6.2. It is the usual sense of weak convergence of probability measures on a topological space that is equipped with the Boreal σ-field, let us here recall it explicitly.

Definition 6.1.5 (Convergence of Random Radon Measures)

A sequence $(\pi_n)_{n \in \mathbb{N}}$ of random Radon measures on D *converges in distribution* towards a random Radon measure π on D if, for any continuous bounded function $F \colon \mathfrak{R}(D) \to \mathbb{R}$, we have $\lim_{n \to \infty} \mathbb{E}[F(\pi_n)] = \mathbb{E}[F(\pi)]$. We then also say that $(\pi_n)_{n \in \mathbb{N}}$ *converges weakly* towards π.

The continuity of F refers of course to the vague topology on $\mathfrak{R}(D)$, which is a priori difficult to characterize. Here is a more handy criterion.

Lemma 6.1.6 (Weak Convergence of Radon Measures in the Vague Topology) *A sequence $(\pi_n)_{n \in \mathbb{N}}$ of random Radon measures on D converges in distribution towards a random Radon measure π if and only if, for any $k \in \mathbb{N}$ and for all relative compact sets $A_1, \ldots, A_k \subset D$ satisfying $\pi(\partial A_i) = 0$ almost surely for all $i \in \{1, \ldots, k\}$, the vector $\big(\pi_n(A_1), \ldots, \pi_n(A_k)\big)_{n \in \mathbb{N}}$ converges weakly towards the vector $\big(\pi(A_1), \ldots, \pi(A_k)\big)$ and can be written $\pi_n \Longrightarrow \pi$.*

Exercise 6.1.7
Prove Lemma 6.1.6 using the characterization in Lemma 2.1.5 and Remark 6.1.3. ◇

---

[2]Prohorov's theorem states that a sequence $(P_n)_{n \in \mathbb{N}}$ of probability measures on $\mathbb{R}^d$ has a weakly convergent subsequence if and only if it is tight. A set Γ of probability measures on $\mathbb{R}^d$ is called *tight* if for any $\varepsilon > 0$ there is a compact set $K \subset \mathbb{R}^d$ such that $P(K^c) < \varepsilon$ for any $P \in \Gamma$.

Another handy criterion is the following. We denote the *Laplace transform* of a random measure π on D by

$$\mathcal{L}_\pi(f) = \mathbb{E}\left[e^{-\int_D f(x)\,\pi(dx)}\right], \qquad f: D \to [0, \infty) \text{ measurable.}$$

Again, this notation is slightly misleading, since $\mathcal{L}_\pi$ is deterministic and does not depend on π but only on its distribution.

Lemma 6.1.8 (Convergence and Laplace Transforms) *A sequence $(\pi_n)_{n\in\mathbb{N}}$ of random Radon measures on D converges in distribution towards a random Radon measure π on D if and only if the Laplace transforms converge, i.e., for any continuous test function $f: D \to [0, \infty)$ with compact support,*

$$\lim_{n\to\infty} \mathcal{L}_{\pi_n}(f) = \mathcal{L}_\pi(f).$$

Exercise 6.1.9
Prove Lemma 6.1.8 using the characterization in Lemma 2.1.6.   ◊

Now we discuss notions of convergence of random point processes that are suitable for the study of ergodic limits as we will consider in ▶ Sect. 6.4. Here we will be concerned with probability measures on the set $\mathbb{S}(\mathbb{R}^d)$ of point clouds in $D = \mathbb{R}^d$, and the limiting probability measure will be again on $\mathbb{S}(\mathbb{R}^d)$. Hence, we will consider only convergence of random point measures towards random point measures. As in earlier chapters, it will be natural to write point measures sometimes as the corresponding sets of points; this will hopefully cause no confusion. We will have later a particular interest in stationary point processes, but this will not be important in this section.

The convergence introduced in Definition 6.1.5 is called in probability theory just the convergence with respect to the weak topology on the Boreal space $\mathfrak{R}(D)$ that is equipped with the vague topology. This is a probability space whose measurable structure comes from a topological structure; its σ-algebra is the one that is generated by the open sets. Definition 6.1.5 can be equivalently formulated by saying that, for probability measures $P, P_1, P_2, P_3, \ldots$ on $\mathfrak{R}(D)$,

$$P_n \Longrightarrow P \quad \Longleftrightarrow \quad \lim_{n\to\infty} \langle F, P_n \rangle = \langle F, P \rangle \quad \text{for any } F \in \mathcal{C}_c(\mathfrak{R}(D)), \qquad (6.1.1)$$

where we write $P_n \Longrightarrow P$ to express that $(P_n)_{n\in\mathbb{N}}$ converges weakly to P. Indeed, if P is the distribution of a random Radon measure π, then $\mathbb{E}[F(\pi)] = \langle F, P \rangle$ for any integrable function F, and (6.1.1) is clearly seen to be a reformulation of Definition 6.1.5.

However, there is a more general notion of weak convergence of probability measures on a general measurable space Ω. Indeed, for a given set $\mathfrak{F}$ of measurable functions $\Omega \to \mathbb{R}$, we introduce, on the set of probability measures under which any $F \in \mathfrak{F}$ is integrable, the notion of *convergence in the weak topology induced by $\mathfrak{F}$* by requiring, for such probability measures $P, P_1, P_2, P_3, \ldots$,

$$P_n \overset{\mathfrak{F}}{\Longrightarrow} P \qquad \Longleftrightarrow \qquad \lim_{n \to \infty} \langle F, P_n \rangle = \langle F, P \rangle \quad \text{for any } F \in \mathfrak{F}. \qquad (6.1.2)$$

Then $\mathfrak{F}$ is called a *convergence-determining set*. If Ω is a Boreal space and $\mathfrak{F} = \mathcal{C}_c(\Omega)$, then (6.1.2) coincides obviously with (6.1.1). If $\mathfrak{F}$ is contained in $\mathcal{C}_c(\Omega)$, then weak convergence implies weak convergence relative to $\mathfrak{F}$. On the other hand, if $\mathcal{C}_c(\Omega)$ is contained in $\mathfrak{F}$, then convergence with respect to $\mathfrak{F}$ implies weak convergence. An example for this is the τ-topology where $\mathfrak{F}$ is given by the set of measurable functions with compact support, see also Remark 2.1.3 but note that there the definition is used for different spaces.

The convergence (6.1.2) is the sense of convergence that we will be relying on in ▶ Sect. 6.4 on the set $\Omega = \mathbb{S}(\mathbb{R}^d \times \mathcal{M})$ of marked point measures for $\mathcal{M}$ a mark space. In fact, we will restrict ourselves to stationary point measures on this Ω, and we will use as a convergence-determining class $\mathfrak{F}$ the set of *local and tame functions* $\mathfrak{T} = \mathfrak{T}(\mathbb{S}(\mathbb{R}^d \times \mathcal{M}))$, see Definition 6.1.10. This test-function class is tailormade for the study of stationary point processes and will play an important role when we consider ergodic theorems for Palm measures (see Lemma 6.7.5) as well as large-deviation principles for Palm measures (see Corollary 7.7.6).

From now on we write $\mathcal{P}(\Omega)$ for the set of probability measures on a Boreal-measurable space Ω. Note that (6.1.2) defines a topology on the set $\mathcal{P}(\Omega)$ which is the coarsest (i.e., smallest) topology that guarantees that all the maps $P \mapsto \langle F, P \rangle$ are continuous for $F \in \mathfrak{F}$.

Let us now introduce and investigate the set of tame local functions. We do this for marked point clouds in $\mathbb{S}(\mathbb{R}^d \times \mathcal{M})$, with $\mathcal{M}$ a measurable mark space as in ▶ Sect. 2.4. It is useful to note that, for functions $g \colon \mathcal{M} \to \mathbb{R}$, counting (functions of) marks in an area can be expressed as

$$\phi(\mathbb{1}_\Lambda \otimes \mathbb{1}) = \sum_{i \in I \,:\, x_i \in \Lambda} g(m_i), \qquad \phi = ((x_i, m_i))_{i \in I}, \, \Lambda \subset \mathbb{R}^d.$$

Definition 6.1.10 (Tame and Local Functions)

We call a measurable function $F \colon \mathbb{S}(\mathbb{R}^d \times \mathcal{M}) \to \mathbb{R}$

1. *local* if it depends only on a compact box $\Lambda \subset \mathbb{R}^d$, i.e., if there is a function $G \colon \mathbb{S}(\Lambda \times \mathcal{M}) \to \mathbb{R}$ such that $F(\phi) = G(\phi \cap (\Lambda \times \mathcal{M}))$ for any $\phi \in \mathbb{S}(\mathbb{R}^d \times \mathcal{M})$,

2. *tame* if it is bounded from above by a constant times one plus the counting functional in some compact box, i.e., if there is a constant $C \in (0, \infty)$ and a compact box $\Lambda \subset \mathbb{R}^d$ such that $|F(\phi)| \leq C(1 + \phi(\mathbb{1}_\Lambda \otimes \mathbb{1}))$ for any $\phi \in \mathbb{S}(\mathbb{R}^d \times \mathcal{M})$.

By $\mathfrak{T} = \mathfrak{T}(\mathbb{S}(\mathbb{R}^d \times \mathcal{M}))$, we denote the set of local tame functions on $\mathbb{S}(\mathbb{R}^d \times \mathcal{M})$.

Remark 6.1.11 (A Variant of Tameness) A variant of the notion of tameness is the following. For a given reference function $\psi \colon \mathcal{M} \to [0, \infty)$, one replaces $\phi(\mathbb{1}_\Lambda \otimes \mathbb{1})$ by $\phi(\mathbb{1}_\Lambda \otimes \psi)$ in the upper bound for tameness, i.e., one requires that $|F(\phi)| \leq C(1 + \phi(\mathbb{1}_\Lambda \otimes \psi))$. This is interesting for applications; see [GeoZes93], which we will revisit in ▸ Sect. 7.7 when we when we discuss error analysis. ◇

Tame local functions do not have to be bounded; they form a natural class of test functions on random point clouds. We will now employ them as a convergence-determining class as in (6.1.2).

Definition 6.1.12 (Tame Topology)

Let $\Phi, \Phi_1, \Phi_2, \ldots$ be random point configurations on $\mathbb{S}(\mathbb{R}^d \times \mathcal{M})$, then we say that Φ_n *converges tamely*, as tamely, as $n \to \infty$, towards Φ if for any tame local function $F \colon \mathbb{S}(\mathbb{R}^d \times \mathcal{M}) \to \mathbb{R}$,

$$\lim_{n \to \infty} \mathbb{E}[F(\Phi_n)] = \mathbb{E}[F(\Phi)].$$

This is (6.1.2) for $\mathfrak{F} = \mathfrak{T}$ rewritten in terms of a sequence of random variables on the same probability space.

A great advantage of the tame topology is that it can be used for a very large class of interesting random point processes.

Exercise 6.1.13

Show that tame local functions form a real vector space. Further show that any tame local function is integrable with respect to any stationary random point process with finite intensity. ◇

Note that every map $\phi = ((x_i, m_i))_{i \in I} \mapsto \langle f, \phi \rangle = \phi(f) = \sum_{i \in I} f(x_i)$ with $f \in \mathcal{C}_c(\mathbb{R}^d)$ is tame. Hence, if probability measures on $\mathbb{S}(\mathbb{R}^d \times \mathcal{M})$ converge tamely, then also with respect to the weak topology based on the vague topology on $\mathbb{S}(\mathbb{R}^d \times \mathcal{M})$, i.e., the weak topology is coarser than the tame one.

Exercise 6.1.14

Show that the tame topology on $\mathcal{P}(\mathbb{S}(\mathbb{R}^d))$ is strictly finer than the weak topology of (6.1.2) based on the vague topology on $\mathbb{S}(\mathbb{R}^d)$ by using the Portmanteau theorem. ◇

6.2   High-Density Limits

Recall the interpretation that we rely on: the points X_i of the random point cloud $\sum_{i \in I} \delta_{X_i}$ are the locations of the devices of the telecommunication system in the communication area $D \subset \mathbb{R}^d$. To simplify things, we now want to assume D to be compact, say a large ball. We want to consider a situation where all the X_i are located in close vicinity to each other, i.e., when we consider a busy area in a city or a large cloud of people attending an event. Then it will not be possible anymore to resolve the effect of any single device in the big cloud. Instead, we are going to find a summarizing way to describe the entire cloud in terms of just one function on D, a density.

In this respect, we sometimes speak of *microscopic* quantities, like each of the devices and its interaction with some other device or with something else, and *macroscopic* quantities, which depend on the entire cloud or large parts of it, i.e., of entities of positive percentages of the devices. This jargon cannot be rigorously defined and depends on the personal perspective and on the limiting setting that one considers, but often it gives a good intuition.

Example 6.2.1 (Transmission to One Base Station)
Imagine that in D there is one base station located at a site in D. Each of the devices X_j of the cloud $(X_i)_{i \in I}$ has a message that wants to be transmitted to the base station. Each such message and the success of its transmission is a microscopic quantity. Often one is interested in the percentage of successfully transmitted messages, which is a macroscopic quantity.

One can make the model more realistic (and more interesting) by adding some or all of the following features.

- The success of a transmission can depend on interference, e.g., in terms of the signal-to-interference ratio defined in ▶ Sect. 5.2. The interference can be based on additional properties, like the signal strength of a device.
- There can be more base stations in the system. A multitude of devices may create interference problems that can be coped with by including additional base stations. Since base stations are expensive, their number will typically be small and can be considered to be of finite order.
- One can admit several hops via the other devices, leading to a *multi-hop system*. We think of hops of lengths of finite order, hence the trajectories will contain only a bounded hop number and are still considered microscopic.

These considerations are fundamental and will be carried out and continued in detail in Example 6.2.4 which is extended in ▶ Sect. 7.4, and in Example 6.2.6, extended in ▶ Sect. 7.5 ◇

Here is the basic kind of limit that we will employ. We consider a PPP and express the high-density situation with one large parameter λ, the order of the number of devices per unit volume.

Lemma 6.2.2 (Convergence of Empirical Measures)
Let $D \subset \mathbb{R}^d$ be compact and μ a measure on D. For $\lambda \in (0, \infty)$, let $\Phi_\lambda = (X_i^{(\lambda)})_{i \in I_\lambda}$ be a PPP in D with intensity measure $\lambda \mu$. Then, as $\lambda \to \infty$, the random point measure

$$L_\lambda = \frac{1}{\lambda} \sum_{i \in I_\lambda} \delta_{X_i^{(\lambda)}} \tag{6.2.1}$$

converges weakly (with respect to the vague topology) towards μ as $\lambda \to \infty$.

Proof

According to Lemma 6.1.8, it is sufficient to check the convergence of the Laplace transform. Let $f: D \to [0, \infty)$ be continuous, then, according to Campbell's theorem (Theorem 2.3.1),

$$\mathcal{L}_{L_\lambda}(f) = \mathcal{L}_{\Phi_\lambda}(f/\lambda) = \exp\left(\int_D (e^{-f(x)/\lambda} - 1)\,\lambda\mu(dx)\right). \qquad (6.2.2)$$

For any x, we see that the integrand with respect to $\mu(dx)$ converges towards $-f(x)$. If f is integrable with respect to μ, then we can apply the dominated convergence theorem, since $1 - e^{-y} \leq y$ for any $y \in \mathbb{R}$, and therefore the integrand is bounded in absolute value by $f(x)$. Hence, we see that the Laplace transform converges towards $\exp(-\int_D f(x)\,\mu(dx)) = \mathcal{L}_\mu(f)$ in this case. If f is not integrable with respect to μ, then we estimate $\mathcal{L}_{L_\lambda}(f)$ first against $\mathcal{L}_{L_\lambda}(f \wedge K)$ for some cutting parameter K, derive convergence towards $\exp\{-\int f \wedge K\,d\mu\}$ and use then the monotone-convergence theorem for letting $K \to \infty$ to see that $\mathcal{L}_{L_\lambda}(f)$ converges towards $0 = \mathcal{L}_\mu(f)$. In both cases, we have verified the convergence of the Laplace transform for all nonnegative continuous test functions f.   □

Exercise 6.2.3

Formulate and prove a version of Lemma 6.2.2 for a CPP Φ_λ with suitable random intensity measure.   ◇

The interpretation of Lemma 6.2.2 is that the dense cloud of devices in D approaches the intensity measure μ, i.e., a multitude of microscopic information is replaced by a much simpler macroscopic object, a measure. For the latter there are good perspectives for further analysis. The high-density limit has the advantage that the limit is rather manageable (it is fully determined by the intensity measure), but the disadvantage that the assumption of many devices in small areas often appears not appropriate (e.g., in daily usual situations without any spatial clumping) and that details of the functionality cannot be resolved on a level that registers every single device.

One might argue that such a limiting setting is useless for describing human beings, since they cannot be squeezed together infinitely strongly. But the most important quantities are quite well described by these asymptotic, and in ▸ Sect. 7.3 we will see that the convergence in Lemma 6.2.2 is even exponentially fast in a certain sense, i.e., the main characteristics of the limiting formula can be clearly seen already for moderate values of λ.

Certainly, in the high-density setting of Lemma 6.2.2 one would like to use also some functionals of the cloud of devices that describe elements of the functionality of the telecommunication system. However, a priori only continuous functionals of the cloud L_λ are amenable to an application of that lemma, but not those that describe single devices. Such functionals will be able only to resolve macroscopic quantities, i.e., objects that depend on a number $\mathcal{O}(\lambda)$ of devices.

Example 6.2.4 (Average Number of Neighbors)

In the setting of Lemma 6.2.2, we fix a length parameter $r \in (0, \infty)$ and call two device locations *neighbors* if their distance is smaller than r. The interpretation is that neighbors can exchange messages directly, which gives a simple telecommunication model that is very much related to the Boolean model that we discuss at length in ▶ Chap. 4.

Consider the number $\Phi_\lambda(B_r(X_i) \cap D)$ of neighbors of $X_i \in \Phi_\lambda$, where $B_r(x)$ denotes the ball around x with radius $r > 0$. For highly dense systems, i.e., for large λ, this number is of order λ. We are interested in the averaged, normalized number of neighbors,

$$M_\lambda = \lambda^{-1} \sum_{i \in I_\lambda} \lambda^{-1} \Phi_\lambda(B_r(X_i) \cap D).$$

This quantity says something about the average number of devices with which a typical device is connected, where the average is taken over all the devices. However, note that the sum on i is not necessarily normalized to one by the factor of $1/\lambda$, like L_λ.

This quantity can be written as a functional of the normalized empirical measure, L_λ. Indeed, observe that, introducing the functional

$$\begin{aligned} G(v) &= \int_D v(\mathrm{d}x) \, v(B_r(x) \cap D) \\ &= \int_{D \times D} v(\mathrm{d}x) v(\mathrm{d}y) \, \mathbb{1}\{\|x - y\| < r\}, \qquad v \in \mathfrak{R}(D), \end{aligned} \tag{6.2.3}$$

we have $M_\lambda = G(L_\lambda)$. It is an exercise to check that G is continuous in the vague topology. Hence, using Lemma 6.2.2 we can deduce that M_λ converges, as $\lambda \to \infty$, to $G(\mu) = \int_D \mu(\mathrm{d}x) \, \mu(B_r(x) \cap D)$.

This example will be further analyzed and extended in ▶ Sect. 7.4. Certainly, it is rather simple and does not have much structure. The proposals that we made in Example 6.2.1 for adding more interesting features apply also here. However, attach to each device a trajectory of a message with some energy would need an extension of Lemma 6.2.2 to marked processes. ◇

Exercise 6.2.5
Prove that the function G defined in (6.2.3) is continuous with respect to the vague topology on $\mathfrak{R}(D)$. ◇

Example 6.2.6 (Connectivity to a Base Station)
We elaborate a bit more on Example 6.2.1. In the situation of Lemma 6.2.2, we assume that D contains the origin o, at which a base station is located. Imagine that each of the devices has a message that is supposed to travel to o, either in a direct hop or in a finite number of hops. Due to interference reasons (see ▶ Chap. 5), not every hop is possible. Denote by $A(X_i, L_\lambda)$, the event that a direct hop from X_i to o is possible, i.e., the transmission of the message $X_i \to o$ is successful. Likewise, denote by $A(X_i, X_j, L_\lambda)$ the event that a two-hop trajectory $X_i \to X_j \to o$ is possible. Such an event for a fixed i is microscopic information and cannot be expressed in terms of the empirical measure L_λ.

Now the interesting quantities are the number of devices that can send directly to o, i.e.,

$$\#\{i \in I_\lambda : A(X_i, L_\lambda) \text{ occurs}\},$$

and the number of devices that can send with at most two hops to o, i.e.,

$$\#\{i \in I_\lambda : A(X_i, X_j, L_\lambda) \text{ occurs for some } j \in I_\lambda\}.$$

Both are macroscopic quantities that can be written as functionals of L_λ. Under quite natural assumptions, they are of order λ. Assuming these functionals (with prefactor $1/\lambda$) are continuous, Lemma 6.2.2 could be used to express the large-λ asymptotic. However, the continuity is not true (at least for some very natural way to define success of a message transmission), and one has to find some technical ways around that problem in order to make precise assertions about this limit. We will discuss this in detail in ▶ Sect. 7.5.   ◊

6.3   Intermezzo: Stationary Processes and Birkhoff's Theorem

In the remaining part of ▶ Chap. 6, we will be concerned with averages of (functionals of) shifts of random point processes over large boxes in $\mathbb{R}^d$. This has much to do with stationarity and ergodic limits. We are going to prepare for that by recalling the well-known situation of stationary processes indexed by $\mathbb{N}$ (interpreted as time), i.e., the discrete, one-dimensional situation. This belongs to the standard material taught in two-semester lectures on probability. This section is meant for discussing fundamental ideas and notions in this situation. In ▶ Sect. 6.4 we will then first give a motivation for the d-dimensional continuous situation and then develop the material independently from scratch.

Let us now talk about averages $n^{-1}(X_1 + \cdots + X_n)$ of integrable random variables $X_1, \ldots, X_n$ in the limit $n \to \infty$. The simplest case is where all these variables are independent and have the same distribution. Here the famous law of large numbers applies and says that the average converges almost surely to the expectation of X_1. However, we aim at handling a much larger class of stochastic processes $(X_n)_{n \in \mathbb{N}}$, the stationary ones. For that we need to introduce the shift operator $\theta \colon \mathbb{R}^\mathbb{N} \to \mathbb{R}^\mathbb{N}$, defined by $\theta((x_i)_{i \in \mathbb{N}}) = (x_{i+1})_{i \in \mathbb{N}}$. We introduce the σ-field $\mathcal{F}_\theta$ as the set of events Γ in the sequence space $\mathbb{R}^\mathbb{N}$ such that $\theta(\Gamma) = \Gamma$, i.e., the set of events that are not changed by shifts in time. This σ-algebra is called the *shift-invariant σ-algebra*. Standard examples of events in $\mathcal{F}_\theta$ are $\{(x_n)_{n \in \mathbb{N}} \colon x_n \in A \text{ for infinitely many } n\}$ or $\{(x_n)_{n \in \mathbb{N}} \colon x_n \in A \text{ for almost all } n\}$ for any measurable set $A \subset \mathbb{R}$.

Here is the fundamental result of long stretch averages of stationary processes.

Theorem 6.3.1 (Birkhoff's Ergodic Theorem)
Let $\Phi = (X_i)_{i \in \mathbb{N}}$ be a stationary sequence of random variables, i.e., the distribution of $\theta(\Phi) = (X_{i+1})_{i \in \mathbb{N}}$ is identical to the one of Φ. Let $f \colon \mathbb{R} \to \mathbb{R}$ be a function such that $f(X_1)$ is integrable. Then $S_n = n^{-1} \sum_{i=1}^{n} f(X_i)$ converges almost surely and in $\mathcal{L}^1$-sense towards $\mathbb{E}[f(X_1)|\mathcal{F}_\theta]$, the conditional expectation of $f(X_1)$ given the σ-field $\mathcal{F}_\theta$ of all shift-invariant events.

Sketch of Proof

We follow the idea of [KatWei82], which seems to work only in the one-dimensional case. By the usual decomposition into positive and negative parts of f, we may assume that f is nonnegative. The limit superior $\overline{f}$ and the limit inferior $\underline{f}$ of $(S_n)_{n \in \mathbb{N}}$ are $\mathcal{F}_\theta$-measurable, as $\overline{f} = \overline{f} \circ \theta$ and $\underline{f} = \underline{f} \circ \theta$. Hence, it suffices to show that $\overline{f} \le \mathbb{E}[f(X_1)]$. Then $\mathbb{E}[f(X_1)] \le \underline{f}$ follows analogously, and we are finished.

The core of the argument is the following. For given $\varepsilon > 0$, we construct inductively a sequence $0 = N_0 < N_1 < N_2 < \dots$ of random indices such that the increments along this subsequence come ε-close to the limit superior, i.e.,

$$f(X_{N_k}) + \dots + f(X_{N_{k+1}-1}) \ge (N_{k+1} - N_k)(\overline{f} - \varepsilon), \qquad k \in \mathbb{N}_0.$$

Here we use the idea that the quotient of the left-hand side and $N_{k+1} - N_k$ is equal to the partial sum $S_{N_{k+1}-N_k}$, applied to $\theta^{N_k}(\Phi)$ instead of Φ. Then we sum all these inequalities over all k such that $N_k \le N$, integrate over the randomness, noting that also the N_k are random, divide by n, let $n \to \infty$ and finally $\varepsilon \downarrow 0$ and are finished. Some technical problems come from the fact that f and the increments $N_{k+1} - N_k$ do not have to be bounded. This is handled by some clever cutting strategy. □

The assertion is a kind of law of large numbers for an average of a measurable quantity over time. It is sometimes roughly summarized by saying that '(limiting) time-average is equal to space-average', at least in the case where the limit is just equal to the deterministic number $\mathbb{E}[f(X_1)]$, see the end of this section for criteria for this. Unlike in the famous law of large numbers for independent and identically distributed random variables, no independence is involved, even though Theorem 6.3.1 contains this situation as a special case.

Remark 6.3.2 (Interpretation of the Ergodic Theorem) An interpretation of Theorem 6.3.1 is the following. Many copies of the random quantity $f(X_i)$ are mixed. They are embedded in a (time-) environment that has the property to be in distribution invariant under shifts, i.e., to show statistically similar long-term strings when shifted by any amount of time. This means that the typical average realization that one sees when looking at the many copies $f(X_i)$ is an average value of $f(X)$, for some copy X of X_1, together with the information that the observed copy is embedded in a long-term environment that is shifted so often that any further shift does not give any new information. This shift-invariance of what I see is the only 'hard' information on what I see. Hence, the typical copy that I see after mixing over many copies is the expectation of $f(X)$ given the event that any shift of the entire sequence is the same event. ◇

Remark 6.3.3 (Measure-Preserving Transformations) The fact that a sequence $\Phi = (X_i)_{i \in \mathbb{N}}$ of random variables is stationary can also be rephrased by saying that its distribution $\mathbb{Q} = \mathbb{P} \circ \Phi^{-1}$ is invariant under the shift-transformation θ, since $\mathbb{P} \circ \Phi^{-1} = \mathbb{P} \circ (\theta(\Phi))^{-1} = \mathbb{P} \circ \Phi^{-1} \circ \theta^{-1}$. Therefore, θ is called a *measure-preserving transformation* on $\mathbb{R}^\mathbb{N}$ for the probability measure $\mathbb{Q}$. We note here that this concept is all that we need for Birkhoff's ergodic theorem, and one finds this theorem in the literature often in a formulation for a general probability space $(\Omega, \mathcal{F}, \mathbb{P})$, equipped with a measure-preserving transformation $T: \Omega \to \Omega$. The assertion is then that, for any integrable random variable $F: \Omega \to \mathbb{R}$, almost surely and in $\mathcal{L}^1$-sense,

$$\lim_{n\to\infty} n^{-1} \sum_{i=1}^{n} F \circ T^i = \mathbb{E}[F|\mathcal{F}_T], \qquad (6.3.1)$$

where $\mathcal{F}_T$ is the σ-algebra of events A in $\mathcal{F}$ that are invariant under T, i.e., $A = T^{-1}(A)$. ◊

Remark 6.3.4 (More General Test Functions) There is a formulation of Birkhoff's ergodic theorem that is – at first sight – much more general than Theorem 6.3.1 by taking far more general test functions f. This is suggested by Remark 6.3.3: why not take a function that depends on the entire sequence $\Phi = (X_i)_{i\in\mathbb{N}}$ instead of only on the first component X_1? Indeed, Theorem 6.3.1 can equivalently also be formulated by saying that, for any measurable function $F: \mathbb{R}^{\mathbb{N}} \to \mathbb{R}$ that is integrable with respect to $\mathbb{Q} = \mathbb{P} \circ \Phi^{-1}$,

$$\lim_{n\to\infty} n^{-1} \sum_{i=1}^{n} F(\theta^i(\Phi)) = \mathbb{E}[F(\Phi)|\mathcal{F}_\theta], \qquad (6.3.2)$$

almost surely and in $\mathcal{L}^1(\mathbb{Q})$-sense. For the special choice $F((x_i)_{i\in\mathbb{N}}) = f(x_1)$, we see immediately that Theorem 6.3.1 is a special case of that. However, using some measure-theoretic standard tools, one sees that both formulations are indeed equivalent, and also equivalent to the formulation in Remark 6.3.3. We keep in mind that the mixture of *every* integrable functional of a stationary stochastic process converges, in particular of those that depend on infinitely many indices. ◊

Remark 6.3.5 (Empirical Measure of Shifts) Let us have a look at the *empirical measure of shifts*,

$$\mathcal{R}_n(\phi) = n^{-1} \sum_{i=1}^{n} \delta_{\theta^i(\phi)}, \qquad \phi = (x_k)_{k\in\mathbb{N}} \in \mathbb{R}^{\mathbb{N}}, \qquad (6.3.3)$$

which is a probability measure on $\mathbb{R}^{\mathbb{N}}$ that registers how many shifts of the given sequence ϕ are in a given subset of $\mathbb{R}^{\mathbb{N}}$. One should observe that

$$n^{-1} \sum_{i=1}^{n} F(\theta^i(\phi)) = \langle F, \mathcal{R}_n(\phi) \rangle, \qquad \phi \in \mathbb{R}^{\mathbb{N}}, F: \mathbb{R}^{\mathbb{N}} \to \mathbb{R},$$

where we write $\langle F, P \rangle$ for $\int F \, dP$. Then Birkhoff's ergodic theorem says that, for any stationary stochastic process Φ, we have

$$\langle F, \mathcal{R}_n(\Phi) \rangle \to \mathbb{E}[F(\Phi)|\mathcal{F}_\theta], \qquad F \in \mathcal{L}^1(\mathbb{Q}),$$

almost surely and in $\mathcal{L}^1(\mathbb{Q})$-sense. ◊

Remark 6.3.6 (Empirical Measures in General) Measures like the one defined in (6.3.3) are often used for describing averaged long-term aspects of stochastic processes replacing all the details of the involved random variables. They contain much information about the involved random variables, but are much easier to handle, since they neglect the order. For example, the empirical measure $L_n = n^{-1} \sum_{i=1}^{n} \delta_{X_i}$ of a random process $(X_i)_{i \in \mathbb{N}}$ registers how often the process hits a given set by time n, since $nL_n(A) = \#\{i \in \{1, \ldots, n\}: X_i \in A\}$. This is the type of measure that we are based on in the high-density setting of ▶ Sect. 6.2.

But much more complex functionals of the process can be sampled and averaged. Empirical k-string measures $n^{-1} \sum_{i=1}^{n} \delta_{(X_i, X_{i+1}, \ldots, X_{i+k-1})}$ register how often a given string of length k appears before time n. The version with $k = \infty$ deals with mixtures of shifts of infinite paths and is particularly useful when dealing with stationary sequences $(X_i)_{i \in \mathbb{N}}$. Certainly, there are d-dimensional versions for random fields indexed by $\mathbb{N}^d$, and the continuous version of that is $\mathcal{R}_W$ in (6.4.7) below. ◇

Let us briefly discuss some properties of the limiting random variable $\mathbb{E}[f(X_1)|\mathcal{F}_\theta]$, in particular the question how random it is. Indeed, in general it is a priori nontrivially random, but there is an important subcase in which it is almost-surely constant, i.e., a deterministic number, more precisely, just $\mathbb{E}[f(X_1)]$. This is the case when Φ has the property that $\mathbb{P}(\Phi \in \Gamma) \in \{0, 1\}$ for each event Γ such that $\theta(\Gamma) = \Gamma$, in other words when $\mathcal{F}_\theta$ is trivial under the distribution of Φ. This property is often called *ergodicity* of Φ in the context of processes indexed by $\mathbb{N}$. This notion is fundamental for stationary processes. i.i.d. sequences Φ are in particular ergodic in this sense, the strong law of large numbers for them is a consequence of Birkhoff's ergodic theorem. Also irreducible Markov chains Φ that have an invariant distribution are ergodic. However, this notion of ergodicity is not very intuitive, very intuitive, nor is it easy to verify in many cases, and its verification often goes via showing the validity of suitable mixing properties.

But we leave now the setting of a discrete, one-dimensional index set and proceed with point processes in $\mathbb{R}^d$.

6.4 Ergodic Theorem for Point Processes

In contrast to the high-density limit discussed in ▶ Sect. 6.2, in this section, we discuss a fundamentally different ansatz for deriving simplifying formulas for highly complex telecommunication systems, i.e., for random point processes with many points. In the remainder of ▶ Chap. 6, we investigate a type of limiting situation where both the considered area and the number of devices are large, and the device intensity, i.e., the expected number of devices per unit volume, is constant. This limiting setting is known from statistical physics as the *thermodynamic limit*. It describes a 'typical' large-space situation, a daily-life situation in a large area without any particularities.

Imagine we are given a complex telecommunication system in a large area $W_n = [-n, n]^d$. It is impossible to gain a comprehensive survey over all the details. But on the other hand, this would not be helpful anyway in order to grasp the main characteristics of the situation. Instead, much more valuable for the understanding of the situation would be some average values of decisive quantities like the portion of devices that are not connected or the portion of devices that are connected to precisely four other devices, or similar information. Such quantities can essentially be written as averages of some observables over W_n. Often the observable is local, i.e., looks only at the configuration in a window whose size does not depend on n, and this window systematically sweeps the area W_n. In the end, we average over what we see in all these shifts of the observation window and would like know what this resembles for large n.

To be a bit more precise, let $\phi = \sum_{i \in I} \delta_{(x_i, m_i)}$ be a marked point cloud in $\mathbb{R}^d$ of points x_i with marks m_i in some mark space $\mathcal{M}$. We try to proceed in the spirit of Birkhoff's ergodic theorem, Theorem 6.3.1. Fix an observation window $\Lambda = [-L/2, L/2]^d$ with some $L > 0$, and assume that our observable $f : \mathbb{S}(\Lambda \times \mathcal{M}) \to \mathbb{R}$ depends only on the configuration in Λ. One tempting way to sweep the observable through $W_n = [-n, n]^d$ is to take n as a multiple of L and to cover W_n by a regular partition given by $(n/L)^d$ windows $Lz + \Lambda$ with $z \in \mathbb{Z}^d$ satisfying $Lz \in W_n$. Then we are interested in the large-n behavior of the average of what f sees in all the L-shifts of the window

$$\mathcal{A}_{n,L}(f, \phi) = \left(\frac{L}{n}\right)^d \sum_{z \in \mathbb{Z}^d : Lz \in W_n} f\left(\theta_{Lz}(\phi) \cap \Lambda\right). \tag{6.4.1}$$

Here the *shift operator* θ_x, for any $x \in \mathbb{R}^d$, is defined by $\theta_x(y) = y - x$. Using the same notation, we then extend the domain of this operator in a standard way to sets $A \subset \mathbb{R}^d$ and have therefore also the shift operator $\theta_x : \mathbb{S}(\mathbb{R}^d) \to \mathbb{S}(\mathbb{R}^d)$. Clearly, θ_x can be further extended to act on subsets $\Gamma \subset \mathbb{S}(\mathbb{R}^d)$.

In the discrete spatial averaging in (6.4.1), the translations of the observation windows are essentially disjoint and cover the entire area. One can see $Lz + \Lambda$ as a 'typical' observation window and $\mathcal{A}_{n,L}(f, \phi)$ as the average of the observable. The good point is that, in the important special case of $\phi = \Phi$ being a stationary PPP, all these observables are independent and identically distributed, and therefore the usual law of large numbers for sums of i.i.d. random variables applies, with the result that

$$\lim_{n \to \infty} \mathcal{A}_{n,L}(f, \Phi) = \mathbb{E}\left[f(\Phi \cap \Lambda)\right], \tag{6.4.2}$$

almost surely and in $\mathcal{L}^1(\mathbb{Q})$. Then the right-hand side of (6.4.2) contains the description of the situation in a 'typical' translation of an observation window, i.e., the relevant local information.

However, there are a number of disadvantages:

(1) If there are non-trivial dependencies between different marks, then the observations in the observation windows $Lz + \Lambda$, $z \in \mathbb{Z}^d$, are not independent.

(2) Interactions between different windows cannot be expressed in terms of $\mathcal{A}_{n,L}(f, \Phi)$.

(3) For many applications, it may be that the choice of one fixed diameter L is not sufficient to describe the situation in satisfactory way.

(4) Applications to more general random point processes cannot be handled in this way.

Drawbacks (1) and (2) could be overcome by taking overlapping windows $L^z_{\frac{1}{2}} + \Lambda$ with $z \in \mathbb{Z}^d$, but then the idea of independence is totally lost. However, having the goal to describe interesting communication areas, i.e., with interesting interdependences between the devices, we are forced to drop independence anyway even though the stationary PPP will be our main source of examples. Instead, the appropriate main model assumption about the random point process Φ under consideration is *stationarity*, i.e., distributional shift-invariance of the entire point cloud, in the same spirit as in Birkhoff's ergodic theorem. Furthermore, we will not consider *discrete* shifts of observation windows, but *continuous* shifts of the entire shifted point process, i.e., shifts by any $x \in \mathbb{R}^d$. This will require some technical work at the boundary of the area W_n, which will not be difficult if the observable is local or well approximated by such observables. Finally, we will also look at shifts by *random* amounts X_i for $i \in I$, which will lead us to the natural notion of Palm measures.

Having given some motivation, we now extend the concept of stationarity and limits of averages from the one-dimensional discrete-parameter setting of ▶ Sect. 6.3 to the continuous d-dimensional setting, i.e., to stationary random point processes in $\mathbb{R}^d$. Hence, the time interval $\{1, 2, 3, \ldots, n\}$ is replaced with a large box in $\mathbb{R}^d$ (often $[-n, n]^d$), and the stochastic process $(X_i)_{i \in \mathbb{N}}$ is replaced with $\sum_{i \in I} \delta_{X_i}$, a stationary point process in $\mathbb{R}^d$. Our main result here will be a version of Theorem 6.3.1 and a discussion of the notion of ergodicity and some examples. See e.g. [MKM78], [DalVer08, Section 12.3] or [GeoZes93] for proofs and more measure-theoretical background. It is good to remember that the intensity measure of a stationary point process with finite intensity is a positive multiple of the Lebesgue measure, see Lemma 2.1.10.

First note that the stationarity of a random point cloud $\Phi = \sum_{i \in I} \delta_{X_i}$ can be rephrased by saying that $\theta_x(\Phi)$ has the same distribution as Φ for any $x \in \mathbb{R}^d$, or by saying that the distribution $\mathbb{Q} = \mathbb{P} \circ \Phi^{-1}$ of the point cloud is invariant under θ_x for any $x \in \mathbb{R}^d$, i.e., $\mathbb{Q} = \mathbb{Q} \circ \theta_x^{-1}$. Again in other words, this means that θ_x is measure-preserving for $\mathbb{Q}$ for any $x \in \mathbb{R}^d$; see Remark 6.3.3. It is important that we have even a *group* $\{\theta_x : x \in \mathbb{R}^d\}$ of measure-preserving transformations, indexed by $\mathbb{R}^d$. Also in this context we denote by $\mathcal{F}_\theta$ the stationary σ-algebra on $\mathbb{S}(\mathbb{R}^d)$ that consists of all the translation-invariant events Γ, i.e., events that are invariant under θ_x for any $x \in \mathbb{R}^d$, i.e., $\Gamma = \theta_x(\Gamma)$ for any x. This is the analog to the σ-algebra $\mathcal{F}_\theta$ appearing in Birkhoff's ergodic theorem. The following set is an example of a translation-invariant event,

$$\{\phi \in \mathbb{S}(\mathbb{R}^d) : \phi(\theta_x(A)) = k \text{ for any } x \in \phi\}, \qquad A \subset \mathbb{R}^d \text{ measurable}, k \in \mathbb{N}_0.$$

Note that there are *two* natural ways to build spatial mixtures of point processes $\sum_{i \in I} \delta_{x_i}$ over large boxes W_n: mixing homogeneously over all shifts by $x \in W_n$, and mixing over all shifts by x_i with $i \in I$ such that $x_i \in W_n$. We start by considering the first setting and come back later to the second one, which is essentially the Palm version of the former.

We now formulate a d-dimensional, continuous-space analog of Birkhoff's ergodic theorem. This will be the central result of this section. For this sake, we need to consider, instead of the time interval $\{1, 2, 3, \ldots, n\}$, a sequence of large areas in space. We call $(W_n)_{n \in \mathbb{N}}$ a *convex averaging sequence* if it is an increasing sequence of convex and compact subsets of $\mathbb{R}^d$ such that $\sup\{r \geq 0 \colon B_r(x) \subset W_n \text{ for some } x\} \to \infty$ as $n \to \infty$. The most natural convex averaging sequence may be $([-n, n]^d)_{n \in \mathbb{N}}$.

Theorem 6.4.1 (Wiener's Ergodic Theorem)

Let $(W_n)_{n \in \mathbb{N}}$ be a convex averaging sequence, and let Φ be a stationary point process in $\mathbb{R}^d$ with distribution $\mathbb{Q} = \mathbb{P} \circ \Phi^{-1}$. Then, for any integrable function $F \colon \mathbb{S}(\mathbb{R}^d) \to \mathbb{R}$, almost surely and in $\mathcal{L}^1(\mathbb{Q})$-sense,

$$\lim_{n \to \infty} |W_n|^{-1} \int_{W_n} F(\theta_x(\Phi)) \, dx = \mathbb{E}[F(\Phi)|\mathcal{F}_\theta]. \tag{6.4.3}$$

The $\mathcal{L}^1(\mathbb{Q})$-convergence in (6.4.3) is called the *mean ergodic theorem*, while the almost sure convergence is called the *individual ergodic theorem*. Wiener's ergodic theorem is an extension of Birkhoff's theorem in the form (6.3.2) in Remark 6.3.4; we took a test function of the entire point process.

Proof

We follow the presentation of the original proof of [Wie39] that is given in [Geo11, Chapter 14.A]. The strategy is the following. We first use Hilbert space arguments and convexity to show, for $F \in \mathcal{L}^2(\mathbb{Q})$, the convergence in $\mathcal{L}^2(\mathbb{Q})$-sense, more precisely, first the convergence of the norms, then the $\mathcal{L}^2(\mathbb{Q})$-convergence. Then we extend the convergence to $\mathcal{L}^1(\mathbb{Q})$-convergence for $F \in \mathcal{L}^1(\mathbb{Q})$, using that $\mathcal{L}^2(\mathbb{Q})$ is dense in $\mathcal{L}^1(\mathbb{Q})$. As a preparation for the proof of the almost-sure convergence, one proves the *maximal ergodic lemma*, which controls the probability that the supremum over n of $|W_n|^{-1} \int_{W_n} F(\theta_x(\Phi)) \, dx$ exceeds a constant, i.e., a tightness result. The final proof of the almost-sure convergence applies the Boreal–Cancelli lemma and is rather technical. Here is the first step:

Lemma 6.4.2 (Minimizers of Convex Sets in Hilbert Spaces)

Let C be a nonempty closed convex set in a Hilbert space $\mathcal{H}$. Then there is a unique $f \in C$ such that $\|f\| = \inf_{g \in C} \|g\|$. Furthermore, any asymptotically minimizing sequence $(f_n)_{n \in \mathbb{N}}$ in C converges towards f.

Proof of Lemma 6.4.2

Put $c = \inf_{g \in C} \|g\|$ and pick some sequence $(f_n)_{n \in \mathbb{N}}$ in C such that $\lim_{n \to \infty} \|f_n\| = c$. In any Hilbert space, the parallelogram identity

$$\frac{1}{4}\|f - g\|^2 = \frac{1}{2}(\|f\|^2 + \|g\|^2) - \frac{1}{4}\|f + g\|^2, \qquad f, g \in \mathcal{H},$$

is satisfied. Hence, for given $\varepsilon > 0$, if $m, n \in \mathbb{N}$ are so large that $\|f_n\| \leq c + \varepsilon$ and $\|f_m\| \leq c + \varepsilon$, then it follows that $\frac{1}{4}\|f_n - f_m\|^2 < \varepsilon$ since $\frac{1}{2}(f_n + f_m) \in C$. Hence $(f_n)_{n \in \mathbb{N}}$ is a Cauchy sequence. Let f be its limit. By continuity, $\|f\| = c$. Since C is closed, $f \in C$. If $g \in C$ with $\|g\| = c$, then again the parallelogram identity gives that $\|f - g\| = 0$. Hence f is unique. $\qquad \square$

Now we assume that $F \in \mathcal{L}^2(\mathbb{Q})$ and show that (6.4.3) holds in the $\mathcal{L}^2(\mathbb{Q})$-sense. We apply Lemma 6.4.2 to $\mathcal{H} = \mathcal{L}^2(\mathbb{Q})$ with norm $\|F\| = \mathbb{E}[F(\Phi)^2]^{1/2}$ and $\mathcal{C}$ equal to the closure of the convex hull of $\{F \circ \theta_x : x \in \mathbb{R}^d\}$. Put $c = \inf_{g \in \mathcal{C}} \|g\|$. We abbreviate

$$R_n(F)(\phi) = |W_n|^{-1} \int_{W_n} F(\theta_x(\phi)) \, dx, \qquad \phi \in \mathbb{S}(\mathbb{R}^d). \tag{6.4.4}$$

First we make evident that it is sufficient to show that $\lim_{n \to \infty} \|R_n(F)\| = c$.

Indeed, Lemma 6.4.2 then implies that $R_n(F)$ converges in $\mathcal{L}^2(\mathbb{Q})$-sense towards the unique element $\overline{F} \in \mathcal{C}$ satisfying $\|\overline{F}\| = c$. Now, for any $x \in \mathbb{R}^d$, $\overline{F} \circ \theta_x$ lies in $\mathcal{C}$ and minimizes the norm (because $\|\overline{F} \circ \theta_x\| = \|\overline{F}\| = c$) and is therefore equal to $\overline{F}$ because of uniqueness of the minimizer. Hence, $\overline{F}$ is $\mathcal{F}_\theta$-measurable. Furthermore, for any $g \in \mathcal{L}^2(\mathbb{Q})$ that is $\mathcal{F}_\theta$-measurable, we have

$$\mathbb{E}\left[(F(\Phi) - \overline{F}(\Phi)) \, g(\Phi)\right] = \lim_{n \to \infty} \mathbb{E}\left[(R_n(F)(\Phi) - \overline{F}(\Phi)) \, g(\Phi)\right]$$

$$\leq \|g\| \lim_{n \to \infty} \|R_n(F) - \overline{F}\| = 0.$$

Hence, $\overline{F}$ satisfies the defining properties of $\mathbb{E}[F(\Phi)|\mathcal{F}_\theta]$. Hence, for showing the $\mathcal{L}^2$-convergence in (6.4.3) it suffices to show that $\lim_{n \to \infty} \|R_n(F)\| = c$ holds.

Let us do that now. Pick $\varepsilon > 0$. By definition of $\mathcal{C}$, we may pick some finite set $\Delta \subset \mathbb{R}^d$ and $t_z \in [0, 1]$ for each $z \in \Delta$ such that $\sum_{z \in \Delta} t_z = 1$ and such that the function $g = \sum_{z \in \Delta} t_z F \circ \theta_z$ satisfies $\|\overline{F} - g\| \leq \varepsilon$. In particular, $\|g\| \leq c + \varepsilon$. Furthermore, for any $n \in \mathbb{N}$,

$$\|R_n(F) - R_n(g)\| = \left\| \sum_{z \in \Delta} t_z [R_n(F) - R_n(F \circ \theta_z)] \right\|$$

$$\leq \sum_{z \in \Delta} t_z \|R_n(F) - R_n(F \circ \theta_z)\|$$

$$= |W_n|^{-1} \sum_{z \in \Delta} t_z \left\| \int_{W_n} F \circ \theta_y \, dy - \int_{W_n + z} F \circ \theta_y \, dy \right\|$$

$$\leq \|F\| |W_n|^{-1} \sum_{z \in \Delta} t_z |W_n \triangle (W_n + z)|,$$

where $\triangle$ denotes the symmetric difference. Since $|W_n \triangle (W_n + z)|$ is of surface order (i.e., $\mathcal{O}(n^{d-1})$), the right-hand side is smaller than ε for all sufficiently large n. For such n we have

$$c \leq \|R_n(F)\| \leq \|R_n(g)\| + \|R_n(F) - R_n(g)\| \leq \|g\| + \varepsilon \leq c + 2\varepsilon,$$

which finishes the proof of $\lim_{n \to \infty} \|R_n(F)\| = c$ and therefore the proof of the $\mathcal{L}^2(\mathbb{Q})$-convergence in (6.4.3) for $F \in \mathcal{L}^2(\mathbb{Q})$.

Now we prove the $\mathcal{L}^1(\mathbb{Q})$-convergence in (6.4.3). First, it is clear from Jensen's inequality and the $\mathcal{L}^2(\mathbb{Q})$-convergence that, for $F \in \mathcal{L}^2(\mathbb{Q})$, the $\mathcal{L}^1(\mathbb{Q})$-convergence holds true as well. The same statement for $F \in \mathcal{L}^1(\mathbb{Q})$ follows, by the fact that $\mathcal{L}^2(\mathbb{Q})$ is dense in $\mathcal{L}^1(\mathbb{Q})$, via the inequalities

$$\mathbb{E}[|R_n(F)(\Phi) - R_n(G)(\Phi)|] \leq \mathbb{E}[|F(\Phi) - G(\Phi)|]$$

and

$$\mathbb{E}\big[|\mathbb{E}[F(\Phi)|\mathcal{F}_\theta] - \mathbb{E}[G(\Phi)|\mathcal{F}_\theta]|\big] \leq \mathbb{E}[|F(\Phi) - G(\Phi)|],$$

which are valid for any $F, G \in \mathcal{L}^1(\mathbb{Q})$. This ends the proof of the $\mathcal{L}^1(\mathbb{Q})$-convergence in (6.4.3).

Now we are heading towards the almost-sure convergence in (6.4.3). The main preparatory step is interesting on its own. We assume here that $(W_n)_{n \in \mathbb{N}}$ is an increasing sequence of centered cubes such that their volumes diverge to infinity.

Lemma 6.4.3 (Maximal Ergodic Lemma)   *For any $F \in \mathcal{L}^1(\mathbb{Q})$ and $c \in (0, \infty)$,*

$$\mathbb{P}\Big(\sup_{n \in \mathbb{N}} |R_n(F)(\Phi)| > c \Big) \leq \frac{3^d}{c} \mathbb{E}[|F(\Phi)|]. \tag{6.4.5}$$

Proof of Lemma 6.4.3

Since $|R_n(F)| \leq R_n(|F|)$, we may and shall assume that $F \geq 0$. Fix $N \in \mathbb{N}$, a discrete cube $\Delta \subset \mathbb{Z}^d$ and a configuration $\phi \in \mathbb{S}(\mathbb{R}^d)$, and consider the set $\Delta_\phi = \{z \in \Delta \colon \theta_z(\phi) \in A_N\}$. For any $z \in \Delta$, there is a radius $n(z, \phi) \in \mathbb{N}$ such that $R_{n(z,\phi)}(F)(\theta_z(\phi)) > c$. Using the notation $U_{z,\phi} = z + \Lambda_{n(z,\phi)}$ where $\Lambda_n = [-n, n]^d$, this inequality may be written as

$$\int_{U_{z,\phi}} F(\theta_x(\phi)) \, \mathrm{d}x > c \, |U_{z,\phi}|. \tag{6.4.6}$$

We now construct recursively sites $z_1, z_2, \ldots, z_l \in \Delta$ such that the sets $U_{z_1,\phi}, \ldots, U_{z_l,\phi}$ are disjoint and cover Δ_ϕ. We write $V_{z,\phi} = z + \Lambda_{2n(z,\phi)}$ and note that the boxes $U_{z,\phi}$ and $V_{z,\phi}$ are concentric, and that $|V_{z,\phi}| = 3^d |U_{z,\phi}|$. Now we begin with $z_1 \in \Delta_\phi$ that maximizes $n(z, \phi)$ over $z \in \Delta_\phi$. Having constructed $z_1, \ldots, z_m$, we pick the $z_{m+1} \in \Delta_\phi \setminus (\bigcup_{i=1}^m V_{z_i,\phi})$ that maximizes $n(z, \phi)$ over all these z. This construction terminates after finitely many steps, as Δ is a finite box. Let $W_\phi = \{z_1, \ldots, z_l\}$. By construction, Δ_ϕ is covered by the (larger) boxes $V_{z_i,\phi}, i \in \{1, \ldots, l\}$, and the (smaller) boxes $U_{z_i,\phi}$ with $i \in \{1, \ldots, l\}$ that are pairwise disjoint. Let U_ϕ denote their union. Then

$$\#\Delta_\phi \leq \sum_{i=1}^l |V_{z_i,\phi}| = 3^d \sum_{i=1}^l |U_{z_i,\phi}|.$$

Summing (6.4.6) over $z \in W_\phi$, we obtain

$$\int_{U_\phi} F(\theta_x(\phi)) \, \mathrm{d}x > c \, |U_\phi|.$$

Since $U_\phi \subset \Delta + \Lambda_N$, we obtain the crucial estimate

$$\frac{3^d}{c} \int_{\Delta+\Lambda_N} F(\theta_x(\phi)) \, dx \geq \#\Delta_\phi = \sum_{z \in \Delta} \mathbb{1}\{\theta_z(\phi) \in A_N\},$$

where A_N is the event

$$A_N = \left\{ \phi : \max_{n=1}^{N} R_n(F)(\phi) > c \right\}$$

Now integrate with respect to ϕ to get

$$\mathbb{P}(\Phi \in A_N)\#\Delta \leq \frac{3^d}{c} |\Delta + \Lambda_N| \mathbb{E}[F(\Phi)].$$

Letting first run Δ through all cofinal cubes and then letting $N \to \infty$, we arrive at the assertion. $\qquad\square$

Now we finish the proof of Wiener's ergodic theorem by showing the almost-sure convergence in (6.4.3). Let $\varepsilon > 0$ be given. Pick a bounded measurable function G with $\mathbb{E}[|F(\Phi) - G(\Phi)|] \leq \varepsilon^2 3^{-d}$. Then we apply the $\mathcal{L}^1(\mathbb{Q})$-convergence in (6.4.3) to obtain an $N \in \mathbb{N}$ such that $\mathbb{E}[|R_n(G)(\Phi) - \mathbb{E}[G(\Phi)|\mathcal{F}_\theta]|] \leq \varepsilon^2 3^{-d}$. By Lemma 6.4.3, we have

$$\mathbb{P}\left(\sup_{n \in \mathbb{N}} |R_n(F - G)(\Phi)| > \varepsilon \right) \leq \varepsilon$$

and

$$\mathbb{P}\left(\sup_{n \in \mathbb{N}} |R_n(R_N(G))(\Phi) - \mathbb{E}[G(\Phi)|\mathcal{F}_\theta]]| \geq \varepsilon \right) \leq \varepsilon.$$

We also have $\mathbb{P}(|\mathbb{E}[F(\Phi) - G(\Phi)|\mathcal{F}_\theta]| > \varepsilon) \leq \varepsilon$. Moreover,

$$\limsup_{n \to \infty} |R_n(R_N(G) - G)(\Phi)|$$

$$\leq |W_N|^{-1} \int_{W_N} \left(\limsup_{n \to \infty} |W_n|^{-1} \int_{(W_n+y)\triangle W_n} G(\theta_x(\Phi)) \, dx \right) dy = 0,$$

since G is bounded. Since $R_n(\mathbb{E}[G(\Phi)|\mathcal{F}_\theta]) = \mathbb{E}[G(\Phi)|\mathcal{F}_\theta]$ almost surely, we conclude that

$$\limsup_{n \to \infty} |R_n(F)(\Phi) - \mathbb{E}[F(\Phi)|\mathcal{F}_\theta]| \leq \limsup_{n \to \infty} |R_n(F - G)(\Phi)|$$

$$+ \limsup_{n \to \infty} |R_n(R_N(G(\Phi) - \mathbb{E}[G(\Phi)|\mathcal{F}_\theta]))| + |\mathbb{E}[(F - G)(\Phi)|\mathcal{F}_\theta]|.$$

Combining all these estimates, we obtain

$$\mathbb{P}\left(\limsup_{n\to\infty} |R_n(F) - \mathbb{E}[F(\Phi)|\mathcal{F}_\theta]| > 3\varepsilon\right) \leq 3\varepsilon.$$

The proof is completed by letting $\varepsilon \downarrow 0$.   □

Remark 6.4.4 (The Empirical Field) Analogously to the Remark 6.3.5, Wiener's ergodic theorem can be equivalently formulated in terms of the convergence of the *empirical field*

$$\mathcal{R}_W(\phi) = |W|^{-1} \int_W \mathrm{d}x \, \delta_{\theta_x(\phi)}, \qquad W \subset \mathbb{R}^d, \phi \in \mathbb{S}(\mathbb{R}^d). \tag{6.4.7}$$

Indeed, Theorem 6.4.1 is equivalent to the statement that

$$\langle F, \mathcal{R}_{W_n}(\Phi)\rangle \to \mathbb{E}[F(\Phi)|\mathcal{F}_\theta], \qquad F \in \mathcal{L}^1(\mathbb{Q}),$$

almost surely and in $\mathcal{L}^1(\mathbb{Q})$-sense.   ◇

As in the discrete, one-dimensional setting, it is of high interest to know something about the limiting random variable $\mathbb{E}[F(\Phi)|\mathcal{F}_\theta]$ that appears in Wiener's ergodic theorem. In particular, the question is interesting whether it is really random or almost-surely deterministic and just equal to the unconditional expected value of $F(\Phi)$. In the discrete one-dimensional case, see the short remark at the end of ▶ Sect. 6.3; this is handled by definition using the notion of ergodicity, but this is far from being easily understandable. For historic reasons, in the d-dimensional continuous setting, one goes therefore via another route, which is more intuitive. We now define the notion of ergodicity that is usually used for stationary point processes, i.e., for stationary $\mathbb{S}(\mathbb{R}^d)$-valued random variables. It is usual to write shifts of a set $\Gamma \subset \mathbb{R}^d$ by $x \in \mathbb{R}^d$ as $\Gamma - x$ rather than $\theta_x(\Gamma)$.

Definition 6.4.5 (Ergodicity)

A stationary point process Φ is called *ergodic* if for any convex averaging sequence $(W_n)_{n\in\mathbb{N}}$ and all measurable sets $\Gamma_1, \Gamma_2 \subset \mathbb{S}(\mathbb{R}^d)$,

$$\lim_{n\to\infty} |W_n|^{-1} \int_{W_n} \mathbb{P}(\Phi \in \Gamma_1 \cap (\Gamma_2 + x))\, \mathrm{d}x = \mathbb{P}(\Phi \in \Gamma_1)\mathbb{P}(\Phi \in \Gamma_2). \tag{6.4.8}$$

This notion of ergodicity is much more intuitive than the one that is preferred for the one-dimensional, discrete version of ▶ Sect. 6.3. In Remark 6.4.10 we will see that they mean the same.

Ergodicity is a property of the random point process with respect to spatial averages in all directions. Indeed, the left-hand side of (6.4.8) is a mixture over all directions in W_n and is therefore a natural continuous and d-dimensional extension of the average that appears in Birkhoff's ergodic theorem. However, this property is often hard to verify directly. The following (stronger) mixing property is a spatial de-correlation property in every individual spatial direction.

Definition 6.4.6 (Mixing)

A stationary point process Φ is called *mixing* if for all measurable sets $\Gamma_1, \Gamma_2 \subset \mathbb{S}(\mathbb{R}^d)$,

$$\lim_{|x|\to\infty} \mathbb{P}(\Phi \in \Gamma_1 \cap (\Gamma_2 + x)) = \mathbb{P}(\Phi \in \Gamma_1)\mathbb{P}(\Phi \in \Gamma_2). \qquad (6.4.9)$$

Exercise 6.4.7 (Mixing $\Longrightarrow$ Ergodic)

Verify via the dominated convergence theorem that mixing point processes are ergodic. $\Diamond$

Exercise 6.4.8 (PPPs Are Mixing)

Prove that the stationary PPP is mixing and therefore ergodic. $\Diamond$

The following notion is the analog to what is called ergodicity in the setting of processes indexed by $\mathbb{N}$.

Definition 6.4.9 (Metric Transitivity)

A stationary point process is called *metrically transitive* if $\mathbb{P}(\Phi \in \Gamma) \in \{0, 1\}$ for all measurable and translation-invariant sets $\Gamma \subset \mathbb{S}(\mathbb{R}^d)$, i.e., if the distribution of Φ is trivial on $\mathcal{F}_\theta$.

The following exercise states the equivalence of metric transitivity and ergodicity. This means, roughly speaking, that ergodic point processes assign probability one or zero to all events that carry no spatial information.

Exercise 6.4.10 (Ergodicity $\Longleftrightarrow$ Metric Transitivity)

Prove that a stationary random point process is ergodic if and only if it is metrically transitive. $\Diamond$

Example 6.4.11 (Translation-Invariant Events)

Let $\Phi = (X_i)_{i \in I}$ be a stationary PPP, then the event

$$\{\Phi(B_r(X_i)) \leq k \text{ for all } X_i \in \Phi\}$$

has probability zero for any $k \in \mathbb{N}_0$ and $r > 0$. On the other hand, the event

$$\{\Phi(B_r(X_i)) = k \text{ for some } X_i \in \Phi\}$$

has probability one for any $k \in \mathbb{N}_0$ and $r > 0$. $\Diamond$

Remark 6.4.12 (Ergodicity of Directing Measures) In view of applications to Cox point processes, CPPs, see ▶ Sect. 3.1, the notion of mixing can also be extended to cover random elements Λ of $\mathcal{M}(D)$, the set of σ-finite measures on D. (Recall that such measures are directing for CPPs.) We have to assume that Λ has a shift-invariant distribution, i.e., that the distribution of Λ is equal to one of $\theta_x(\Lambda) = \Lambda \circ \theta_x^{-1} = \Lambda + x$ for any $x \in \mathbb{R}^d$. In this case, one calls Λ mixing if, for all bounded measurable functions f, g on $\mathcal{M}(D)$,

$$\lim_{|x| \to \infty} \mathbb{E}[f(\Lambda)g(\Lambda + x)] = \mathbb{E}[f(\Lambda)]\mathbb{E}[g(\Lambda)].$$

Ergodicity of Λ is defined accordingly. For the corresponding properties of the directed CPP, see Exercise 6.4.14. ◊

Exercise 6.4.13 (Stabilizing $\Longrightarrow$ Mixing)
Show that any stabilizing random measure (see ▶ Sect. 3.1) with shift-invariant distribution is mixing in the sense of Remark 6.4.12. ◊

Exercise 6.4.14 (Ergodicity of CPPs)
Prove that a CPP with ergodic directing measure (in the sense of Remark 6.4.12) is ergodic. ◊

Exercise 6.4.15 (Non-ergodicity of Point Processes)
Prove that the mixed PPP of Exercise 3.1.3, i.e., the CPP on $\mathbb{R}^d$ with intensity measure $\Lambda = Z dx$, where Z is a nonnegative random variable, is not ergodic if its directing measure is truly random. ◊

Exercise 6.4.16 (Randomly Shifted Grid)
Let Y be a random variable that is uniformly distributed over $[0, 1]^d$. Prove that the point process $\theta_Y(\mathbb{Z}^d)$ is stationary and metrically transitive and hence ergodic, but not mixing.

Exercise 6.4.17 (Superposition of Ergodic Point Processes)
Let Φ and Ψ be two independent ergodic point processes. Prove that the superposition $\Phi \cup \Psi$ is also ergodic if additionally one of the two processes is mixing. Find a counterexample if the assumption of mixing is dropped. (The solution is in [Kre85, Theorem 3.6].) ◊

6.5   Applications in Telecommunications

Let $\Phi = (X_i)_{i \in I}$ be a stationary point process with distribution $\mathbb{Q}$ and let $W_n = [-n, n]^d$. In telecommunications, one is often interested in quantities of the form

$$A_{W_n}^{(g)}(\Phi) = \sum_{i \in I \,:\, X_i \in W_n} g\big(\theta_{X_i}(\Phi) \cap W_n\big), \tag{6.5.1}$$

where $g \colon \mathbb{S}(\mathbb{R}^d) \to \mathbb{R}$ is measurable. This quantity counts a certain function of those particles in the window W_n that have a certain property that depends on the particles around, but only inside W_n. Examples that we have in mind are of the form $g(\phi) = \#\{x \in \phi : o \text{ has a connection to } x\}$ if on ϕ there is a message transmission rule defined, possibly with the help of a signal-dependence like the SINR of ▶ Sect. 5.2, or the Boolean model of ▶ Sect. 4.3. See Examples 6.5.5 and 6.6.7.

It is tempting to try to apply Wiener's ergodic theorem to $\mathcal{A}^{(g)}_{W_n}(\Phi)$ and to find the existence and a formula for $\lim_{n \to \infty} |W_n|^{-1} \mathcal{A}^{(g)}_{W_n}(\Phi)$. A direct application is not possible, since $\mathcal{A}^{(g)}_{W_n}(\Phi)$ is not of the form $\int_{W_n} F(\theta_x(\Phi)) \, dx$ for any F. Indeed, $\mathcal{A}^{(g)}_{W_n}(\Phi)$ is a *point average* (i.e., an average over the points of the random process) rather than a *continuous shift average*. However, under suitable conditions on g and Φ, it can be approximated in such a way. The main requirement will be that g is a local and tame function (see Definition 6.1.10) and that Φ has a finite intensity. Then, for some suitable F, the difference between $\mathcal{A}^{(g)}_{W_n}(\Phi)$ and $\int_{W_n} F(\theta_x(\Phi)) \, dx$ will consist only of boundary terms, whose influence vanishes in the limit as n tends to infinity. In this section, we explain how this works.

The function F in Wiener's ergodic theorem is sometimes called an *observable*. One good candidate of an observable for handling the quantity $\mathcal{A}^{(g)}_{W_n}(\Phi)$ in (6.5.1) is

$$F_g(\phi) = \sum_{i \in I : x_i \in U} g(\theta_{x_i}(\phi)), \qquad \phi = (x_i)_{i \in I} \in \mathbb{S}(\mathbb{R}^d),$$

where $U = [-\frac{1}{2}, \frac{1}{2}]^d$ is the centered unit box. Indeed, in the limit as $n \to \infty$ we will approximate as follows,

$$\mathcal{A}^{(g)}_{W_n}(\Phi) \sim \sum_{i \in I : X_i \in W_n} g(\theta_{X_i}(\Phi)) \sim \int_{W_n} F_g(\theta_x(\Phi)) \, dx \sim |W_n| \mathbb{E}[F_g(\Phi) | \mathcal{F}_\theta].$$

$$(6.5.2)$$

The result will be the following.

Proposition 6.5.1 (Ergodic Theorem for Point Averages) *Let* Φ *be a stationary point process in* $\mathbb{R}^d$ *with distribution* $\mathbb{Q}$ *and intensity measure equal to a constant multiple of the Lebesgue measure. Let* $g \colon \mathbb{S}(\mathbb{R}^d) \to \mathbb{R}$ *be local and tame. Then, for* $W_n = [-n, n]^d$,

$$\lim_{n \to \infty} |W_n|^{-1} \mathcal{A}^{(g)}_{W_n}(\Phi) = \mathbb{E}[F_g(\Phi) | \mathcal{F}_\theta] \qquad \text{in } \mathcal{L}^1(\mathbb{Q})\text{-sense.} \qquad (6.5.3)$$

As we will explain in Remark 6.5.2, this is a first version of the ergodic theorem for Palm measures; a second one will appear in Theorem 6.7.7.

The proof consists of making the three steps in (6.5.2) rigorous. The last one directly follows from Wiener's ergodic theorem, as soon as one has understood that F_g is integrable, which is a minor extension of Exercise 6.1.13. In the following Remark 6.5.3, we explain the second step for the special case $g \equiv 1$, and we defer the general case to Exercise 6.5.4. Proving the first step in (6.5.2) is quite similar; we do not give details.

Remark 6.5.2 (Connection to Palm Versions) Let us have a short look at Proposition 6.5.1 in the light of Palm theory. We restrict to the case where the limit is constant, i.e., where it is equal to

$$\mathbb{E}[F_g(\Phi)] = \mathbb{E}\Big[\sum_{i \in I \, : \, X_i \in U} g\big(\theta_{X_i}(\Phi)\big)\Big].$$

An obvious observation is that this is equal to the Palm expectation of g, i.e., to $\lambda\mathbb{E}[g(\Phi^*)]$, where we recall from ▶ Sect. 2.5 that Φ^* is the Palm version of Φ, and the normalizing constant is $\lambda|U| = \lambda$, where we assume that the intensity measure of Φ is $\lambda\,\mathrm{d}x$.

The fact that the limit is naturally written in terms of a Palm expectation suggests that there might be an alternative route of proving Proposition 6.5.1. This is indeed true, and we will describe this way in the last part of ▶ Sect. 6.6, in the more complex setting where we add random marks to the points of Φ. An outline of this strategy is the following.

1. Write $\mathcal{A}_{W_n}^{(g)}(\Phi)$ as an integral with respect to the empirical field $\mathcal{R}_{W_n}(\Phi)$ defined in Remark 6.4.4.
2. Show that the Palm measure of $\mathcal{R}_{W_n}(\Phi)$ is asymptotically for $n \to \infty$ well approximated by the empirical individual field $|W_n|^{-1} \sum_{i \in I \, : \, X_i \in W_n} \delta_{\theta_{X_i}}(\Phi)$.
3. Show that taking Palm measures is a continuous operation in suitable topologies.
4. Deduce a Palm version of Wiener's ergodic theorem using this map.   ◊

Proposition 6.5.1 is useful for local and tame observables g. However, many observables are not local, but can be approximated with spatial cut-off versions of such observables by taking a monotone limit. As a consequence, for many applications, it will be sufficient to work with tame functions and their monotone limits, which has to be carried out on a case-by-case basis. A relevant example of an observable that cannot be approximated with finite-window versions appears at the end of Example 6.5.5.

Note that counting observables g of *pairs* of points is in general not tame but may be integrable, see Example 6.5.6, and a version of Proposition 6.5.1 can be proved as well.

Wiener's ergodic theorem is well-suitable for introducing a convincing notion of micro- and macroscopic objects in the limit as $n \to \infty$. A local observable F is called *microscopic*, since it depends only on a bounded part of the point process, and the integral $\int_{W_n} F(\theta_x(\Phi))\,\mathrm{d}x$ is called *macroscopic*, since it describes a global property of the random point process.

The following gives a proof for the second step in (6.5.2) for $g = 1$ and is therefore a building stone of the proof of Proposition 6.5.1. It relies on the control of boundary effects for counting observables.

Remark 6.5.3 (Counting Particles in Large Boxes) It is illustrative and instrumental to have a short look at a *local* test function F of the form $F(\phi) = F((x_i)_{i \in I}) = \phi(\Lambda) = \#\{i \in I : x_i \in \Lambda\}$, the number of points in a finite box $\Lambda \subset \mathbb{R}^d$. Recall from Exercise 6.1.13 that $F(\Phi)$ is is integrable if the intensity measure of Φ is a multiple of the Lebesgue measure, which we want to assume here. Hence Wiener's ergodic theorem says that the quantity

$\int_{W_n} F(\theta_x(\Phi)) \, dx$ behaves like $|W_n|$ times the conditional expected number of points of Φ in Λ given $\mathcal{F}_\theta$. This quantity is given by

$$\int_{W_n} F(\theta_x(\Phi)) \, dx = \int_{W_n} \#\{i \in I : X_i \in \Lambda + x\} \, dx$$
$$= \sum_{i \in I} |\{x \in W_n : x \in \Lambda - X_i\}| = \sum_{i \in I} |W_n \cap (\Lambda - X_i)|. \tag{6.5.4}$$

Note that this quantity is not equal but close to $|\Lambda| \Phi(W_n)$, to the volume of Λ times the number of points in W_n. Moreover, it depends on more of the points of Φ than only on those in W_n, namely on all the points in $W_n + \Lambda$. Indeed, if $\Lambda = [-\frac{L}{2}, \frac{L}{2}]^d$ and and $W_n = [-n, n]^d$, then the random variables $\Phi(W_n)$ and $\int_{W_n} F(\theta_x(\Phi)) \, dx$ differ only in the boundary set $\Delta_{n,L} = [-n - L, n + L]^d \setminus [-n + L, n - L]^d$. More precisely,

$$\left| \int_{W_n} F(\theta_x(\Phi)) \, dx - |\Lambda| \Phi(W_n) \right| \leq \sum_{i \in I : X_i \in \Delta_{n,L}} \left(|W_n \cap (\Lambda - X_i)| + |\Lambda| \right)$$

$$\leq 2|\Lambda| \Phi(\Delta_{n,L}),$$

since $|W_n \cap (\Lambda - x)| = |\Lambda|$ if $x \in [-n+L, n-L]^d$. It can be easily shown that the right-hand side is $o(|W_n|)$ as $n \to \infty$ in probability, since the volume of $\Delta_{n,L}$ is of order n^{d-1}. Hence, the difference between $\int_{W_n} \theta_x(\Phi)(\Lambda) \, dx$ and $|\Lambda| \Phi(W_n)$ comes only from a boundary effect. ◊

We keep in mind the rule of thumb that counting points with certain properties in a given box W_n cannot easily be expressed in terms of the integral appearing on the left-hand side of (6.4.3), but for local properties the difference comes only from boundary effects, which vanish in the limit $n \to \infty$, after dividing by the volume of W_n.

The next exercise deals with the extension of Remark 6.5.3 from constant functions to local tame functions. This is the main step in the proof of Proposition 6.5.1.

Exercise 6.5.4 (Counting Functionals in Large Boxes)
Prove the second step in (6.5.2) in the setting of Proposition 6.5.1. That is, assume that Φ is a stationary point process with intensity measure equal to a multiple of the Lebesgue measure, and assume that $g : \mathbb{S}(\mathbb{R}^d) \to \mathbb{R}$ is local and tame. Then, as $n \to \infty$,

$$A_n^{(g)}(\Phi) \sim \int_{W_n} F_g(\theta_x(\Phi)) \, dx \qquad \text{in } \mathcal{L}^1(\mathbb{Q})\text{-sense}.$$

Hint: In the difference between $A_n^{(g)}(\Phi)$ and $\int_{W_n} F_g(\theta_x(\Phi)) \, dx$, write everything under one sum on $i \in I$ (interchanging it with $\int_{W_n} dx$ for the second term), and distinguish the cases $X_i \in W_{n-1}$ and the rest. Observe that the first term is zero, and estimate the expectation of the absolute values of the two other terms with the help of the locality and tameness of g. Use that the sum on $i \in I$ can be restricted to those i such that $X_i \in \Lambda + (W_{n+1} \setminus W_{n-1})$ for some box Λ. ◊

Example 6.5.5 (Connectivity)

Let us have a look at the Boolean model, which we discussed in ▶ Chap. 4, in the light of the ergodic theorem. We fix a PPP $\Phi = (X_i)_{i \in I}$ in $\mathbb{R}^d$ with intensity measure λdx, and we draw a connection from X_i to X_j if their distance is positive but smaller than some number $R \in (0, \infty)$. Then one might be interested in the proportion of the number of points X_i in the window $W_n = [-n, n]^d$ that are directly connected to at least one other device in W_n, i.e., are not isolated in the Poisson–Gilbert graph in W_n. Note that this quantity depends on W_n in two ways, it counts in W_n, and the connection must be to some point in W_n. It is easily seen that the number we are after is equal to $\mathcal{A}^{(g)}_{W_n}(\Phi)$ defined in (6.5.1) with g given as

$$g(\phi) = \mathbb{1}\{o \text{ is not isolated}\} = \mathbb{1}\{\phi(B_R(o) \setminus \{o\}) \geq 1\}.$$

This is obviously local and tame. Hence, the observable

$$F(\phi) = F_g(\phi) = F_g((x_i)_{i \in I}) = \#\{i \in I : x_i \in U, x_i \text{ is not isolated}\},$$

where $U = [-\frac{1}{2}, \frac{1}{2}]^d$ can be used in Proposition 6.5.1, and we obtain that (6.5.3) holds. Remark that the limit $\mathbb{E}[F(\Phi)|\mathcal{F}_\theta]$ is not random, but equal to $\mathbb{E}[F(\Phi)]$, since the PPP Φ is ergodic (see Exercise 6.4.8) and therefore metrically transitive (see Definition 6.4.9 and Exercise 6.4.10) and therefore $\mathcal{F}_\theta$ is trivial. Hence, we get that

$$\lim_{n \to \infty} |W_n|^{-1} \#\{i \in I : X_i \text{ is not isolated in } W_n\} = \mathbb{E}\big[\#\{\text{non-isolated points in } U\}\big].$$

Another interesting quantity is the number of points X_i in W_n that can reach at least $l \in \mathbb{N}$ other points by at most $k \in \mathbb{N}$ hops in the Gilbert graph. This is the quantity $\mathcal{A}^{(g)}_{W_n}(\Phi)$ with g given by

$$g(\phi) = \mathbb{1}\big\{\#\{i \in I : o \overset{k}{\leftrightsquigarrow} x_i\} \geq l\big\}.$$

Here we recall the notation for k-hop connectivity,

$$x \overset{k}{\leftrightsquigarrow} y \quad \Longleftrightarrow \quad \text{there are } i_1, \ldots, i_{k-1} \in I$$

$$\text{such that } 0 \leq |X_{i_{m-1}} - X_{i_m}| < R, m = 1, \ldots, k, \quad (6.5.5)$$

with the convention that $x = X_{i_0}$ and $y = X_{i_k}$. Note that the indices $i_1, \ldots, i_k$ do not have to be pairwise different. Also g is a local and tame function, as is easily verified. Hence, we have from Proposition 6.5.1 and the above remark on ergodicity that (6.5.3) holds, i.e., $\lim_{n \to \infty} |W_n|^{-1} \mathcal{A}^{(g)}_{W_n}(\Phi) = \mathbb{E}[F_g(\Phi)]$.

However, the number of devices in W_n that are connected to *infinitely* many other devices with arbitrarily many hops (i.e., that lie in the giant component $\mathcal{C}$ of the Poisson–Gilbert graph based on Φ and R) is not microscopic, and the corresponding functional is not local. Indeed, it depends by its very nature on the entire point process in $\mathbb{R}^d$, not only in W_n. However, it can be formulated in terms of the observable

$$F(\phi) = \#\{i \in I : x_i \in U \cap \mathcal{C}\},$$

which is obviously integrable. Hence, Wiener's ergodic theorem (Theorem 6.4.1) is applicable. Interestingly, we can identify the limit as follows,

$$\lim_{n \to \infty} |W_n|^{-1} \#\{i \in I : X_i \in W_n \cap \mathcal{C}\} = \mathbb{E}[F(\Phi)] = \lambda P^o(o \in \mathcal{C}) = \lambda \theta(\lambda, R),$$

where $\theta(\lambda, R)$ is the percolation probability from (4.3.2), and we use Campbell's formula. $\Diamond$

Here is an example of an integrable observable that is not tame in general.

Remark 6.5.6 (Counting Point Pairs) It is sometimes of interest to find an asymptotic formula for the number of pairs of points in a large box that satisfy some local condition. Examples are the number of neighboring pairs in a Boolean model, i.e., the number of pairs that can directly exchange messages, or the number of pairs that are k-hop connected with each other in the Poisson–Gilbert graph. Consider an observable that counts all pairs (X_i, X_j) in the random point cloud that satisfy this local condition and such that X_i lies in U, the unit cube. This observable is local, but a priori not tame, since no other natural upper bound is visible than the product of $\Phi(U)$ and $\Phi(\Lambda)$, where Λ is a sufficiently large box such that it can be decided within Λ whether or not the condition is satisfied. However, such an upper bound is not sufficient to deduce tameness, hence a direct application of Proposition 6.5.1 is not possible. One way out of this is to use just Wiener's theorem (which does not use tameness, but only integrability) and to handle the boundary terms with techniques like the ones that we demonstrated in Remark 6.5.3. This works if the random variable $\Phi(U)\Phi(\Lambda)$ is integrable, which is the case for the standard PPP.

However, it may be that the point process Φ does not make $\Phi(U)\Phi(\Lambda)$ integrable. Or it may be that one definitely needs tameness for continuity issues (e.g, when performing limits in the tame topology). In these cases, one should try to approximate with cutting variables of the form $\Phi(U)K\mathbb{1}\{\Phi(\Lambda) \leq K\}$ with $K \to \infty$, a technique that may also sometimes be successful if the condition is not local. $\Diamond$

6.6 Ergodic Theorem for Marked Point Processes

In this section, we extend Wiener's ergodic theorem, Theorem 6.4.1, to point processes *with marks*, which is obviously necessary to obtain interesting applications, some of which we give later in this section. We will see that all we have to do is to introduce and understand the concept of markings of random point processes and to realize that the proof that we gave for Wiener's ergodic theorem applies also here. The ergodic theorem for stationary point processes with marks is often attributed to Tempel'man, see [DalVer08, Section 12.2].

We fix a mark space $(\mathcal{M}, \mathcal{B}(\mathcal{M}))$ as in ▶ Sect. 2.4, i.e., we assume it to be locally compact. We have discussed only marked *Poisson* point processes, so some few remarks on general ones are in order. Generally, a random marked point process in $\mathbb{R}^d$ is nothing but an $\mathbb{S}(\mathbb{R}^d \times \mathcal{M})$-valued random variable, but recall from the beginning of ▶ Sect. 2.4 that the condition of having pairwise distinct points in a configuration $\phi = ((x_i, m_i))_{i \in I}$ is restricted to the points x_i, i.e., in a configuration ϕ there may be many points having the same mark. In ▶ Sect. 2.4 we encountered mainly random marked point processes $((X_i, M_i))_{i \in I}$ that have a particular probabilistic structure: the random point process $(X_i)_{i \in I}$ is a PPP with intensity measure μ, and given this process, the marks M_i are attached independently over i to X_i, with a distribution that possibly depends on X_i. The assumption that $(X_i)_{i \in I}$ is a PPP can easily be dropped here, but we will go even further and will in general not assume anything about the stochastic dependence structure. Our main aspect will be just stationarity with respect to the first coordinate in $((X_i, M_i))_{i \in I}$.

Let us be precise here. Even though a random marked point process can be seen as an $\mathbb{S}(\mathbb{R}^d \times \mathcal{M})$-valued random variable, its ergodic structure will entirely go back to the one of $\mathbb{R}^d$, which is another aspect in which $\mathbb{R}^d$ and $\mathcal{M}$ do not play analogous roles. Indeed, the shift operator θ_x acts only on the spatial part, not on the mark part. Explicitly, $\theta_x((y, m)) = (y - x, m)$ for $(y, m) \in \mathbb{R}^d \times \mathcal{M}$, and $\theta_x(\phi) = \theta_x(\sum_{i \in I} \delta_{(x_i, m_i)}) = \sum_{i \in I} \delta_{(x_i - x, m_i)}$ for $\phi \in \mathbb{S}(\mathbb{R}^d \times \mathcal{M})$ and so on. The shift-invariant σ-algebra $\mathcal{F}_\theta$ is the σ-field of all events Γ in $\mathbb{S}(\mathbb{R}^d \times \mathcal{M})$ that are shift-invariant in the sense that $\Gamma = \theta_x(\Gamma)$ for any $x \in \mathbb{R}^d$. We will call a random marked point process $((X_i, M_i))_{i \in I}$ *stationary* if its distribution $\mathbb{Q}$ is invariant under every θ_x, i.e., if the distributions of $((X_i, M_i))_{i \in I}$ and $((X_i - x, M_i))_{i \in I}$ are identical for any x. Note that this enforces, if the marking is an independent one, that their distributions are not allowed to depend on the location where they are attached to. For example, if we consider a K-MPPP in the sense of Definition 2.4.1, then the probability kernel K is not allowed to depend on its first argument. As in the unmarked case, we can associate to any stationary marked point process an *intensity* $\lambda \in [0, \infty]$ given by the expected number of unmarked points in a unit volume, i.e., $\lambda = \mathbb{E}[\Phi([-1/2, 1/2]^d \times \mathcal{M})]$.

Now we can formulate the main result of this part of the section. For the rest of the section, $(W_n)_{n \in \mathbb{N}}$ denotes a convex averaging sequence, see the definition prior to Theorem 6.4.1.

Theorem 6.6.1 (Ergodic Theorem for Stationary Marked Point Processes)
Let $\Phi = \sum_{i \in I} \delta_{(X_i, M_i)}$ be a stationary marked point process, i.e., an $\mathbb{S}(\mathbb{R}^d \times \mathcal{M})$-valued random variable whose distribution $\mathbb{Q}$ is invariant under θ_x for any $x \in \mathbb{R}^d$. Then, for any test function $F \in \mathcal{L}^1(\mathbb{Q})$,

$$\lim_{n \to \infty} |W_n|^{-1} \int_{W_n} F(\theta_x(\Phi)) \, dx = \mathbb{E}[F(\Phi)|\mathcal{F}_\theta], \tag{6.6.1}$$

almost surely and in $\mathcal{L}^1(\mathbb{Q})$-sense.

Proof

In short, an inspection of the proof of Theorem 6.4.1 shows that we can simply replace $\mathbb{S}(\mathbb{R}^d)$ by $\mathbb{S}(\mathbb{R}^d \times \mathcal{M})$ and correspondingly replace the measures $\mathbb{Q}$. Let us recall some of the details. If $\mathbb{Q}$ is a probability measure on $\mathbb{S}(\mathbb{R}^d)$ that is invariant under any θ_x with $x \in \mathbb{R}^d$, where $(\theta_x)_{x \in \mathbb{R}^d}$ is a group of measurable maps $\theta_x \colon \mathbb{S}(\mathbb{R}^d) \to \mathbb{S}(\mathbb{R}^d)$, then, for any $f \in \mathcal{L}^1(\mathbb{Q})$, we have that $\lim_{n \to \infty} |W_n|^{-1} \int_{W_n} f \circ \theta_x \, dx = \mathbb{E}[f | \mathcal{F}_\theta]$, where $\mathcal{F}_\theta$ is the set of all measurable sets in $\mathbb{S}(\mathbb{R}^d)$ that are invariant under any θ_x. We quickly check if we may apply this to the present situation, i.e., to $S(\mathbb{R}^d)$ replaced by $\mathbb{S}(\mathbb{R}^d \times \mathcal{M})$, $\mathbb{Q}$ now a probability measure on $\mathbb{S}(\mathbb{R}^d \times \mathcal{M})$, the group $(\theta_x)_{x \in \mathbb{R}^d}$ as defined prior to the theorem, and $f = F(\Phi)$. Indeed, the stationarity of Φ means that $\theta_x(\Phi) = \sum_{i \in I} \delta_{(X_i - x, M_i)}$ has the same distribution as Φ for any $x \in \mathbb{R}^d$, hence $\mathbb{Q}$ is still invariant under any θ_x. Then one has to understand that the random variable Φ is to be conceived as the identity map on $\mathbb{S}(\mathbb{R}^d \times \mathcal{M})$, and then we have $f \circ \theta_x = F(\Phi) \circ \theta_x = F(\theta_x(\Phi))$. Hence, the proof of Theorem 6.4.1 covers also Theorem 6.6.1. $\qquad\square$

As we pointed out in Remark 6.4.4 for the unmarked case, Theorem 6.6.1 can also be reformulated by saying that the empirical field $\mathcal{R}_{W_n}(\Phi)$ converges almost surely and in $\mathcal{L}^1(\mathbb{Q})$-sense towards the conditional distribution of Φ given $\mathcal{F}_\theta$, and the convergence is in the same sense as is explained there. The definition of the empirical field is literally the same, but one must keep in mind that there are marks attached to the points of the process.

Having realized that marking is only a minor extension as it concerns the ergodic theorem, we turn now to the same question about ergodicity. First observe that all the related notions of ergodicity, mixing and metric transitive that we introduced in ▶ Sect. 6.4 directly extend to random marked point processes, as they refer only to spatial shifts, i.e., only to the application of the map θ_x acting on the first component. Furthermore, it can easily be inspected that all assertions about the relations between these notions also extend to the marked case. In particular, we can make the observation that the limit on the right-hand side of (6.6.1) is almost surely constant if Φ is ergodic.

Remark 6.6.2 (Loss of Ergodicity by Marking) It may easily be that ergodic unmarked point processes lose ergodicity when we attach marks, even if the marking is stationary. Examples are easily found by making the marking highly dependent. For example, if all points of a stationary PPP in $\mathbb{R}^d$ are jointly marked by zeros with probability $p \in (0, 1)$ and by ones otherwise, then the event $\{\lim_{n \to \infty} \#\{\text{marks} = 0 \text{ in } W_n\}/|W_n| = 0\}$ is shift-invariant, but has probability p. $\qquad\diamond$

Let us complement Remark 6.6.2 with a positive result that states that i.i.d. markings preserve ergodicity.

Exercise 6.6.3 (I.i.d. Marking Preserves Ergodicity)
Prove that equipping an ergodic unmarked point process with i.i.d. marks produces an ergodic marked point process. The same statement is true if 'ergodic' is replaced by 'mixing'. *Hint*: It suffices to consider the limit as $x \to \infty$ for $\mathbb{P}(\Phi \in \Gamma_1, \Phi \in \Gamma_2 + x)$ for events of the form $\Gamma_k = \{\#\{i \in I \colon X_i \in A_k, M_i \in B_k\} = n_k\}$ for $k \in \{1, 2\}$ with compact sets $A_1, A_2 \subset \mathbb{R}^d$,

measurable sets $B_1, B_2 \in \mathcal{M}$ and $n_1, n_2 \in \mathbb{N}_0$. Condition on $(X_i)_{i \in I}$ and use that, for $|x|$ sufficiently large, the used indices are distinct and therefore the marks are independent conditioned on conditioned on $(X_i)_{i \in I}$.   ◇

In particular, every independently marked stationary PPP $\Phi = ((X_i, M_i))_{i \in I}$ is ergodic. Hence the limit in (6.6.1) is deterministic. Most of the following examples for telecommunication applications are of this form.

Example 6.6.4 (Boolean Model with Random Radii)
Let the independent mark $M_i \in (0, \infty)$ of the point X_i be the (random) radius of the communication area around X_i, i.e., the device at X_i can directly transmit messages only within the ball $B_{M_i}(X_i)$. This extends the Boolean model (see Examples 6.2.4 and 6.5.5) to random radii. The observables considered in Example 6.5.5 are certainly as interesting as before, and the applications of the ergodic theorem that we gave in Example 6.5.5 can also be made in the present situation of random radii, but one must be aware of the fact that the limiting expectations also extend over the radii.   ◇

Example 6.6.5 (SIR with Random Transmission Powers)
A popular modeling element is an attachment of an individual random power (or a random transmission radius) to each of the transmitters in a spatial random telecommunication system. It makes sense to pick them in an i.i.d. way, given the transmitters. Assuming that $(X_i)_{i \in I}$ is a stationary point process and that each transmitter X_i sends its message with signal strength (power) P_i, we then obtain an independently marked stationary point process $\Phi = ((X_i, P_i))_{i \in I}$, see Definition 2.4.1. Recall that letting the independent marks depend on the location would destroy the stationarity. If no interference is registered, then it may be more handy to replace P_i by the radius R_i of the ball around X_i inside which X_i can successfully transmit, this leads to Example 6.6.4.

In the case where we do work with interference as a criterion for success of a transmission, the power P_i will typically appear in the signal-to-interference-ratio defined in (5.2.1). Then one might ask, for example, for any information that concerns the coverage zone of X_i, i.e., the set

$$C_i = \left\{ y \in \mathbb{R}^d : \mathrm{SINR}(X_i, y, ((X_j, P_j))_{j \in I}) \geq \tau \right\} - X_i, \qquad i \in I.$$

This set C_i contains a lot of information about the connectivity of X_i. One of the questions that can be handled using Theorem 6.6.1 is the average number of transmitters in a large box whose coverage zone lies in a certain set A of subsets of $\mathbb{R}^d$, i.e.,

$$F_W(\Phi) = \#\{i \in I : X_i \in W, C_i \in A\}.$$

Interesting choices of A are the set of those sets that contain a centered ball with a fixed given radius, or $A = \{\emptyset\}$. Now, $F_{W_n}(\Phi)$, the number of transmitters X_i in $W_n = [-n, n]^d$ such that $C_i \in A$ is, up to boundary effects, equal to $\int_{W_n} F_U(\theta_x(\Phi)) \, dx$, with $U = [-\frac{1}{2}, \frac{1}{2}]^d$ the unit box, as we explained in Remark 6.5.3 for a similar example. However, note that the asymptotic control of the boundary terms needs here some more care and work if the path-loss function ℓ used in the definition of the SINR does not have bounded support. Hence, Theorem 6.6.1 yields that this number is asymptotically equal to $|W_n|\mathbb{E}[F_U(\Phi)|\mathcal{F}_\theta]$ as $n \to \infty$. ◇

Remark 6.6.6 (Transmitters and Receivers) Like in Example 6.6.7, we sometimes consider pairs of two independent random point processes Φ and Ψ, the process of the transmitters and the one of the receivers, jointly on one probability space and having a lot of interaction. Using marks, we can easily join the two processes to a superposition $\Phi \cup \Psi$ and mark the points in Φ with a 't' and the ones in Ψ with an 'r'. By this construction, we obtain a random marked point process with mark space $\mathcal{M} = \{t, r\}$. If Φ and Ψ are stationary, then the marked superposition is stationary as well. However, as we saw in Exercise 6.4.17, it is in general not true that it is ergodic, even if both Φ and Ψ are ergodic. Furthermore, in general it is also difficult to determine the distribution of the marks; they may have long-reaching dependencies of many kinds. Hence, an autonomous description of the distribution of the marked superposition is in general a difficult problem.

There is one important special case in which the distribution of the superposition is easily understood. This is the case where Φ and Ψ are independent stationary PPPs with intensities λ and μ, respectively. Their marked superposition is another stationary PPP with intensity $\lambda + \mu$ and independent Bernoulli-distributed marks 't' and 'r' with parameter $\lambda/(\lambda + \mu)$. It is clear that we can retrieve the pair (Φ, Ψ) in distribution from this MPPP by taking Φ as the random point process of the points with mark 't' and Ψ, the remaining ones. ◇

Example 6.6.7 (Connectivity to Base Stations and Bounded-Hop Percolation)
As in Example 6.5.5, we are given two independent stationary PPPs $\Phi = (X_i)_{i \in I} \subset \mathbb{R}^d$ and $\Psi \subset \mathbb{R}^d$ with parameters λ and μ, respectively. From each point $X_i \in \Phi$, a message is transmitted that goes via at most k hops over points of Φ in the in the Poisson–Gilbert graph with parameter R, until it finally reaches a point of Ψ. Hence, the points in Φ are the transmitters, and the ones in Ψ are the receivers. The messages hop only along bonds with at least one of the two ends belonging to Φ, and the trajectory ends at the first arrival at a site in Ψ. The idea is that Ψ is the set of base stations, and their connections are wired and have infinite capacity of unbounded lengths, which do not appear in this model.

We assume that all the message transmissions (better: transmission attempts) are carried out at the same time instant. This respects the rule that, at any time instant, a given device can only transmit or receive, but not both. In general, the superposition $\Phi \cup \Psi$ is the set of locations of all the devices of the system, but it is divided into transmitters and receivers only for the time instant that we consider. According to Remark 6.6.6, conceive now the superposition of Φ and Ψ as ONE PPP $\widetilde{\Phi}$ with marks 't' and 'r' for 'transmitters' and 'receivers'. We We conceive $\Phi \subset \widetilde{\Phi}$ as the set of 't'-marked points and $\Psi \subset \widetilde{\Phi}$ as the set of 'q'-marked points.

Our goal is to introduce an entire-space version of the model called 'bounded-hop percolation' in ► Sect. 4.7 (with $\mu = 1/r^d$ in the notation there) and to show that the latter actually arises via the ergodic theorem from the one that we are discussing now.

We draw a line between any two transmitters of Φ and a line between any pair of a transmitter and a receiver if their distance lies in $(0, R)$. With a fixed parameter $k \in \mathbb{N}$, we call a point in $\mathbb{R}^d$ k-hop connected to Ψ if there is a trajectory along the bonds within Φ with at most k hops, arriving at some point in Ψ. We then write $x \overset{k,\Phi}{\leftrightsquigarrow} \Psi$. This is the model that we called *bounded-hop percolation* in ▶ Sect. 4.7. There we studied it in the unit box U and examined some properties as a function depending on λ and k.

Let us see what knowledge the ergodic theorem can contribute. We will show that the unit box quantities of ▶ Sect. 4.7 actually appear as large box averages. We ask about the number of devices in $W_n = [-n, n]^d$ that are k-hop connected with some base station. Some care is necessary to determine this quantity precisely, since we have to fix whether or not the base station needs to lie in W_n and whether or not the connecting devices need to lie in W_n, but it is clear from Proposition 6.5.1, in particular from Remark 6.5.3, that the differences between all these variants of the definition are negligible in the limit $n \to \infty$, as the notion of k-hop connectivity is local. Using the notation in (4.7.1) for the k-hop coverage zone Ξ_k, we are therefore now talking about the observable

$$F_{k,R,U}(\phi, \psi) = \#\{i \in I : x_i \in U \cap \phi, x_i \overset{k,\phi}{\leftrightsquigarrow} \psi\}$$

$$= \#\{i \in I : x_i \in U \cap \Xi_k\}$$

$$= \#(U \cap \phi \cap \Xi_k),$$

where we recall that $U = [-\frac{1}{2}, \frac{1}{2}]^d$ is the unit cube. Note that this observable is equal to F_g for some suitable local function g, after extension to the pair (superposition) (ϕ, ψ), i.e., to marked processes.

Then Theorem 6.6.1, together with a control of the boundary terms as in Remark 6.5.3, gives that

$$\lim_{n \to \infty} |W_n|^{-1} F_{k,R,W_n}(\Phi, \Psi) = \lim_{n \to \infty} |W_n|^{-1} \#(W_n \cap \Phi \cap \Xi_k) = \mathbb{E}[F_{k,R,U}(\Phi, \Psi)]$$

$$= \lambda \Theta_k,$$

where we recall from ▶ Sect. 4.7 that Θ_k is equal to λ^{-1} times the expected number of k-hop connected devices in the unit cube. That is, the quantity Θ_k that was examined there is the corresponding limiting spatial average in large boxes.

One interpretation of the set Ψ is that of the set of base stations. It is tempting to assume that it is not random, but given as the deterministic grid $\Psi = r\mathbb{Z}^d$, since one could stipulate that the operator can make a deterministic choice of where to place the base stations in order to achieve a good coverage (while the locations of the devices in Φ are really random). Then the joint system $\Phi \cup \Psi$ has a distributional shift invariance by any vector in $r\mathbb{Z}^d$ only, i.e., it is not stationary in the sense that we are using, but stationary with respect to a discrete group of shifts, $\{\theta_{rz} : z \in \mathbb{Z}^d\}$. In this case, one can use a discrete, d-dimensional version of Wiener's ergodic theorem and the appropriate notion of shift invariance, to deduce the proper limit for the same quantity. Another possibility is to shift the lattice $r\mathbb{Z}^d$ by a uniformly distributed vector in rU, as in Exercise 6.4.16, to enforce continuous shift invariance.

This type of example will be discussed further in ▶ Sect. 6.8.   ◊

6.7 Empirical Stationary and Individual Fields

In this section, we extend the marked version of Wiener's ergodic theorem, Theorem 6.6.1, to the stationary empirical field in the tame topology and to the individual empirical field, which counts points of the process that have a certain property. The first extension can also be seen as a restriction, since we use only tame local test functions. Its proof only needs to control boundary effects: it is a routine extension in the spirit of Remark 6.4.4. The latter extension is rather desirable for telecommunication applications, as we explained at the beginning of ▶ Sect. 6.5. Proving the ergodic theorem for the individual field will rely on showing that it is equal to the Palm version of the stationary field and that taking Palm measures is continuous in the tame topology. This makes the proof rather transparent and understandable. This route will also turn out helpful in ▶ Sect. 7.7, when we discuss frustration probabilities, i.e., probabilities of very improbable events with bad connectivity properties.

Let us proceed with the extension of Theorem 6.6.1 to a variant of the empirical field $\mathcal{R}_W(\phi)$ defined in Remark 6.4.4 that is even stationary for any ϕ, i.e., it is a member of the set $\mathcal{P}_\theta(\mathbb{S}(\mathbb{R}^d \times \mathcal{M}))$ of stationary marked point clouds. Fix a marked point cloud $\phi = \sum_{i \in I} \delta_{(x_i, m_i)} \in \mathbb{S}(\mathbb{R}^d \times \mathcal{M})$ and, for a centered box $W = [-L/2, L/2]^d$, we let $\phi^{(W)}$ denote the point configuration that is obtained from ϕ by repeating the configuration in W at any integer shift of W. In formulas,

$$\phi^{(W)} = \sum_{z \in \mathbb{Z}^d} \sum_{i \in I : x_i \in W} \delta_{(x_i + Lz, m_i)}. \tag{6.7.1}$$

Then the measure

$$\mathcal{R}_W^{(s)}(\phi) = |W|^{-1} \int_W \delta_{\theta_x(\phi^{(W)})} \, dx \in \mathcal{P}_\theta(\mathbb{S}(\mathbb{R}^d \times \mathcal{M})), \qquad \phi \in \mathbb{S}(\mathbb{R}^d \times \mathcal{M}), \tag{6.7.2}$$

is indeed (the distribution of) a stationary marked point process, which is easily checked as an exercise. It is called the *empirical stationary field*. Actually, one can say that $\mathcal{R}_W^{(s)}(\phi)$ arises from $\mathcal{R}_W(\phi)$ by imposing periodic boundary conditions, a familiar trick to obtain less cumbersome formulas in connection with boundary issues. Let us mention as another well-known example the empirical pair measure $n^{-1} \sum_{i=1}^n \delta_{(X_{i-1}, X_i)}$ of a Markov chain $(X_i)_{i \in \mathbb{N}}$, whose left and right marginal measures are automatically equal to each other if we replace X_0 by X_n.

We saw already in Remark 6.5.3 and Exercise 6.5.4 how to control boundary terms of large boxes. Let us point out what test functions allow for neat formulas when integrated against the empirical stationary field.

Remark 6.7.1 (Counting Points Using the Empirical Stationary Field) We integrate the test function $N_\Lambda^{(f)}(\Phi) = \Phi(\mathbb{1}_\Lambda \otimes f) = \sum_{i \in I : X_i \in \Lambda} f(M_i)$ where $\Lambda \subset \mathbb{R}^d$ is any box and $f : \mathcal{M} \to \mathbb{R}$ any measurable function of the mark space. Then, for any marked cloud

$\phi = \sum_{i \in I} \delta_{(x_i, m_i)} \in \mathbb{S}(\mathbb{R}^d \times \mathcal{M})$, and for any centered box W with volume > 1, we have the following identity,

$$|W|^{-1} N_W^{(f)}(\phi) = \langle \mathcal{N}_U^{(f)}, R_W^{(s)}(\phi) \rangle, \tag{6.7.3}$$

where we write U for the unit box $[-1/2, 1/2]^d$. The proof of this statement is tedious and therefore omitted here, it can be found in [GeoZes93, Lemma 2.3(1)].

Equation (6.7.3) is particularly interesting if we consider the limit $W \uparrow \mathbb{R}^d$. If the function $N_U^{(f)}$ belongs to the convergence determining set, i.e., if the map $P \mapsto \langle P, N_U^{(f)} \rangle$ is continuous, then the ergodic theorem, Theorem 6.6.1, gives that $|W|^{-1} N_W^{(f)}(\Phi)$ converges towards $\mathbb{E}[N_U^{(f)}(\Phi)] = \mathbb{E}[\sum_{i \in I : X_i \in U} f(M_i)]$ if $\Phi = \sum_{i \in I} \delta_{(X_i, M_i)}$ is a stationary marked PPP, for example. ◇

The stationary empirical field and the empirical field are close to each other at least with respect to test integrals against local tame functions, as the next exercise shows.

Exercise 6.7.2 (The Empirical Field Is Asymptotically Stationary)
Let Φ be a stationary point process, and let $W_n = [-n, n]^d$. Show that asymptotically, as $n \to \infty$, the empirical field $\mathcal{R}_{W_n}(\Phi)$ and its stationary version, $\mathcal{R}_{W_n}^{(s)}(\Phi)$, are close to each other in $\mathcal{L}^1(\mathbb{Q})$-sense in the tame topology, that is, for any local and tame test function F,

$$\lim_{n \to \infty} \mathbb{E}\Big[\big| \langle \mathcal{R}_{W_n}(\Phi), F \rangle - \langle \mathcal{R}_{W_n}^{(s)}(\Phi), F \rangle \big| \Big] = 0.$$

See [GeoZes93, Remark 2.4] for a sketch of the proof. ◇

As a consequence, the empirical stationary field satisfies the spatial mean ergodic theorem in the tame topology:

Corollary 6.7.3 (Ergodic Theorem for the Empirical Stationary Field) *Let Φ be an ergodic marked point process with distribution $\mathbb{Q}$, and let $W_n = [-n, n]^d$. Then $\mathcal{R}_{W_n}^{(s)}(\Phi)$ converges tamely towards Φ, i.e., for any local and tame test function F,*

$$\lim_{n \to \infty} \langle \mathcal{R}_{W_n}^{(s)}(\Phi), F \rangle = \mathbb{E}[F(\Phi)] \qquad in \ \mathcal{L}^1(\mathbb{Q}).$$

Now we proceed with another extension of the marked version of Wiener's ergodic theorem, namely to another natural version of a spatial averaging, the one that averages shifts by the points of the process itself over a large box W. We are talking about the measure

$$\mathcal{R}_W^{(s), o}(\phi) = \frac{1}{|W|} \sum_{i \in I : x_i \in W} \delta_{(m_i, \theta_{x_i}(\phi^{(W)}))} \in \mathcal{P}(\mathcal{M} \times \mathbb{S}(\mathbb{R}^d \times \mathcal{M})),$$

$$\phi = ((x_i, m_i))_{i \in I} \in \mathbb{S}(\mathbb{R}^d \times \mathcal{M}), \tag{6.7.4}$$

which is often referred to as the *empirical individual field*. We recall from (6.7.1) that W is a centered box in $\mathbb{R}^d$ and $\phi^{(W)}$ is the point configuration that is obtained from $\phi = ((x_i, m_i))_{i \in I}$ by repeating the configuration in W at any translation of W by an integer vector. Note that, for any $x \in \phi \cap W$, the cloud $\theta_x(\phi^{(W)})$ has a point at the origin, and its Dirac measure is summed over every such point in W. That is, $\mathcal{R}_W^{(s),o}(\phi)$ sees the point cloud $\phi^{(W)}$ from each of the devices in W and mixes all these views. Areas with higher density of devices appear more often in a given observation window. Roughly speaking, it is this measure (and not $\mathcal{R}_W$) that gives direct information about the percentage of devices in W that have a certain property. Hence, $\mathcal{R}_W^{(s),o}(\phi)$ is of quite some interest in view of telecommunication applications.

Note that the normalization $|W|^{-1}$ does not turn $\mathcal{R}_W^{(s),o}(\phi)$ into a probability measure. A proper normalization would be given by the number of points $x_i \in W$ of $\phi = ((x_i, m_i))_{i \in I}$ without regard to their marks. In the random setting considered below, this normalization will appear as an expected number of points of ϕ in the unit volume, its intensity. We will then see in Remark 6.7.6 below that $\mathcal{R}_W^{(s),o}(\phi)$ is equal to the Palm measure of the empirical stationary field $\mathcal{R}_W^{(s)}(\phi)$ defined in (6.7.2), where we use the unnormalized version of the Palm measure first introduced in Theorem 2.5.1. Since there we considered only unmarked processes, we give here some details following [GeoZes93]. It is a natural extension of the Palm concept, but some care is needed to handle the mark of the particle at the origin.

For any $P \in \mathcal{P}_\theta(\mathbb{S}(\mathbb{R}^d \times \mathcal{M}))$, there is precisely one finite measure $P^o \in \mathcal{P}(\mathcal{M} \times \mathbb{S}(\mathbb{R}^d \times \mathcal{M}))$, the *Palm measure* of P, that satisfies

$$
\int_{\mathbb{S}(\mathbb{R}^d \times \mathcal{M})} \int_{\mathbb{R}^d \times \mathcal{M}} f(x, m, \theta_x(\phi))\, \phi(dx, dm)\, P(d\phi)
$$
$$
= \int_{\mathcal{M} \times \mathbb{S}(\mathbb{R}^d \times \mathcal{M})} \int_{\mathbb{R}^d} f(x, m, \phi)\, dx\, P^o(dm, d\phi),
$$

(6.7.5)

for any nonnegative measurable test function f on $\mathbb{R}^d \times \mathcal{M} \times \mathbb{S}(\mathbb{R}^d \times \mathcal{M})$. In particular,

$$
\langle g, P^o \rangle = \int_{\mathbb{S}(\mathbb{R}^d \times \mathcal{M})} \int_{\mathbb{R}^d \times \mathcal{M}} \mathbb{1}\{x \in C\} g(m, \theta_x(\phi))\, \phi(dx, dm)\, P(d\phi),
$$

(6.7.6)

for any nonnegative measurable function g on $\mathcal{M} \times \mathbb{S}(\mathbb{R}^d \times \mathcal{M})$, where $C \subset \mathbb{R}^d$ is measurable with $|C| = 1$. The marginal measure $\mu_P(\cdot) = P^o(\cdot \times \mathbb{S}(\mathbb{R}^d \times \mathcal{M}))$ is called the *mark intensity measure*. It satisfies the identity $\mu_P(B) = E[\Phi([-1/2, 1/2]^d \times B)]$ for any measurable set B in $\mathcal{M}$. In particular, the total mass of P^o is equal to $P^o(\mathcal{M} \times \mathbb{S}(\mathbb{R}^d \times \mathcal{M})) = \mu_P(\mathcal{M})$, which equals precisely the intensity λ of P as defined in the beginning of the section. Please note that compared to the Palm measure introduced in Theorem 2.5.1 we are not normalizing here by $\lambda = \mu_P(\mathcal{M})$. This has the advantage that we can directly identify Palm versions of stationary empirical fields as individual empirical fields, see below.

Exercise 6.7.4

Derive and prove a version of the inversion formula for Palm calculus, see Theorem 3.3.1, for marked random point processes. ◊

Let us consider the map $P \mapsto P^o$. We equip $\mathcal{P}_\theta(\mathbb{S}(\mathbb{R}^d \times \mathcal{M}))$ with the tame topology (see Definition 6.1.10), and we introduce on $\mathcal{P}(\mathcal{M} \times \mathbb{S}(\mathbb{R}^d \times \mathcal{M}))$ the weak topology that is induced by the test integrals against all local bounded functions, similarly to (6.1.2).[3] This topology is called the *local topology* and it corresponds to the τ-topology introduced in Remark 2.1.3.

Lemma 6.7.5 (Taking Palm Measures Is Tamely Continuous) *The map $P \mapsto P^o$ is affine (i.e., $(aP_1 + (1-a)P_2)^o = aP_1^o + (1-a)P_1^o$ for any $a \in [0, 1]$ and any $P_1, P_2 \in \mathcal{P}_\theta(\mathbb{S}(\mathbb{R}^d \times \mathcal{M}))$), injective and continuous with respect to the tame and local topologies.*

Proof

It follows directly from (6.7.5) that the map is affine. If C is bounded and g lies in the convergence-determining set for the considered topology on $\mathcal{P}(\mathcal{M} \times \mathbb{S}(\mathbb{R}^d \times \mathcal{M}))$ (i.e., is local and bounded), then the map $\phi \mapsto \int_{\mathbb{R}^d \times \mathcal{M}} \mathbb{1}\{x \in C\} g(m, \theta_x(\phi)) \phi(\mathrm{d}x, \mathrm{d}m)$ in the inner integral on the right-hand side of (6.7.6) lies in $\mathfrak{T}$, the set of local tame functions. This shows that the map $P \mapsto P^o$ is continuous. From (6.7.5) it can be deduced (see [DalVer03, Lemma 12.1.III]) that P can be uniquely recovered from P^o, i.e., that map is also injective. □

Remark 6.7.6 (Palm Measure of the Empirical Stationary Field) For any $\phi \in \mathbb{S}(\mathbb{R}^d \times \mathcal{M})$ and any box $W \subset \mathbb{R}^d$, the empirical individual field is equal to the Palm measure of the empirical stationary field, that is,

$$[\mathcal{R}_W^{(s)}(\phi)]^o = \mathcal{R}_W^{(s),o}(\phi).$$

This assertion is shown by checking (6.7.5). Indeed (see also [GeoZes93, Remark 2.3]), for any measurable nonnegative function f and $W = [-L/2, L/2]^d$, we have that

$$|W| \int_{\mathbb{S}(\mathbb{R}^d \times \mathcal{M})} \int_{\mathbb{R}^d \times \mathcal{M}} f(x, m, \theta_x(\zeta)) \, \zeta(\mathrm{d}x, \mathrm{d}m) \, \mathcal{R}_W^{(s)}(\phi)(\mathrm{d}\zeta)$$

$$= \int_W \int_{\mathbb{R}^d \times \mathcal{M}} f(x - y, m, \theta_x(\phi^{(W)})) \, \phi^{(W)}(\mathrm{d}x, \mathrm{d}m) \, \mathrm{d}y$$

$$= \sum_{z \in \mathbb{Z}^d} \int_W \int_{\mathbb{R}^d \times \mathcal{M}} f(x - Lz - y, m, \theta_{x+Lz}(\phi^{(W)})) \, \phi(\mathrm{d}x, \mathrm{d}m) \, \mathrm{d}y$$

---

[3]If we would proceed according to Remark 6.1.11 and incorporate a reference function $\psi : \mathcal{M} \to [0, \infty)$ in the definition of tameness, then one should consider all local functions that are bounded in absolute value against a constant multiplied by the function $(x, m) \mapsto \psi(m)$.

$$= \int_{\mathbb{R}^d} \int_{\mathbb{R}^d \times \mathcal{M}} f(x - y, m, \theta_x(\phi^{(W)})) \, \phi(dx, dm) \, dy$$

$$= \int_{\mathbb{R}^d \times \mathcal{M}} \int_{\mathbb{R}^d} f(y, m, \theta_x(\phi^{(W)})) \, dy \, \phi(dx, dm)$$

$$= |W| \int_{\mathcal{M} \times \mathbb{S}(\mathbb{R}^d \times \mathcal{M})} \int_{\mathbb{R}^d} f(y, m, \zeta) \, dy \, \mathcal{R}_W^{(s),o}(\phi)(dm, d\zeta),$$

where in the third line we use that $\theta_{x+Lz}(\phi^{(W)}) = \theta_x(\phi^{(W)})$, and in the fourth line we use that Lebesgue integration is invariant under reflections. Hence, (6.7.5) has been verified. ◇

Now we formulate the main result on the empirical individual field of this section, an ergodic theorem in the local bounded topology if the underlying point process is ergodic.

Theorem 6.7.7 (Ergodic Theorem for Empirical Individual Fields)

Let $\Phi = ((X_i, M_i))_{i \in I}$ be an ergodic point process in $\mathbb{S}(\mathbb{R}^d \times \mathcal{M})$ with distribution $\mathbb{Q}$ and let $W_n = [-n, n]^d$. Then, in $\mathcal{L}^1(\mathbb{Q})$-sense, as $n \to \infty$, $\mathcal{R}_{W_n}^{(s),o}(\Phi)$ converges in the local topology towards the Palm measure $\mathbb{Q}^o$. That is, for any local and bounded measurable function $G \colon \mathcal{M} \times \mathbb{S}(\mathbb{R}^d \times \mathcal{M}) \to \mathbb{R}$, in $\mathcal{L}^1(\mathbb{Q})$-sense,

$$\lim_{n \to \infty} \frac{1}{|W_n|} \sum_{i \in I \colon X_i \in W_n} G\big(M_i, \theta_{X_i}(\Phi^{(W_n)})\big) = \int_{\mathcal{M} \times \mathbb{S}(\mathbb{R}^d)} G(m, \phi) \, \mathbb{Q}^o(dm, d\phi).$$

$$(6.7.7)$$

The proof is simply done by pointing out that $\mathcal{R}_{W_n}^{(s)}(\Phi)$ satisfies the ergodic theorem in the tame topology by Corollary 6.7.3 and using Lemma 6.7.5 and Remark 6.7.6. It is this simplicity that made us choose this route via the empirical stationary measure and then its Palm version: we have now a complete proof for the ergodic theory in the local topology, which applies to a lot of interesting examples. We will enjoy another nice advantage of this restriction to the tame topology in ▶ Sect. 7.7, when we derive a large deviations principle for $\mathcal{R}_{W_n}^{(s),o}$ in the same simple way from such a principle for $\mathcal{R}_{W_n}^{(s)}$. See [DalVer08] for much more general versions of Theorem 6.7.7 (with involved proofs) that take into account also non-ergodic point processes. In this case the definition of the Palm version requires the existence of versions of conditional expectations given the σ-algebra $\mathcal{F}_\theta$. On the other hand, the assumption on the test functions G is then just integrability.

Remark 6.7.8 (Empirical Mark Field) If we take test functions G in (6.7.7) that do not depend on the second argument, then we consider the projection on the first component, the mark space. In other words, we obtain an ergodic theorem for the *empirical mark field*, $|W_n|^{-1} \sum_{i \in I \colon X_i \in W_n} \delta_{M_i}$, which is a random measure on the mark space alone. Like the empirical individual field, it is not normalized. However, if $(X_i)_{i \in I}$ is a stationary PPP with intensity λ and if the marks are i.i.d. random variables (conditional given the point process $(X_i)_{i\in I}$), then the expected total mass of the empirical mark field is equal to λ. Indeed, it is then equal to an empirical measure of a Poisson number with parameter $\lambda|W_n|$ of i.i.d. random variables, divided by $|W_n|$. The ergodic theorem is then a simple consequence of the law of large numbers for the mixture of many i.i.d. random variables.   ◇

6.8   Describing Connectivity with Empirical Individual Fields

We now discuss a modeling ansatz for handling typical questions for several versions of a fundamental type of a spatial telecommunication system. We describe how to bring the ergodic theorem for empirical individual fields, Theorem 6.7.7, into action in order to study some relevant large-scale behavior.

Imagine $(X_i)_{i\in I}$ and $(Y_j)_{j\in J}$ are two independent stationary point processes in $\mathbb{R}^d$. The most standard assumption would be that they are standard PPPs, but much more general processes are possible, and we do not want to restrict ourselves too early. Each X_i has a message that would like to be transmitted to any of the Y_j. That is, $(X_i)_{i\in I}$ is the random point process of the transmitters, and $(Y_j)_{j\in J}$ the random point process of the receivers. All the message transmission attempts are carried out simultaneously at one time instant. Let us assume that the success of a given transmission attempt is defined in some way, see the discussion below. We are interested in a number of large-scale questions like the following.

1. What percent of the transmitters are successful, i.e., how many have at least one Y_j to which the transmission is successful?
2. What percent of the transmitters can successfully transmit to at least, or precisely, a given number N of receivers?
3. What percent of the transmitters can successfully transmit to at least, or precisely, a given number of receivers within, or outside, a certain distance?

More precisely, we are asking about the quotient of the number of successful points X_i in the large box $W_n = [-n, n]^d$ and their entire cardinality in W_n. We would be also happy if in the denominator there is just the volume of W_n, since we can easily evaluate the limit of the quotient of the total number of transmitters in W_n and $|W_n|$. Hence, we are heading towards integrals of the empirical individual field with respect to some functionals. However, it is not clear a priori what a good ansatz for the underlying random point process is.

All these questions and more can be handled in terms of the following random marked point process. Let $\Psi^{(i)}$ denote the subset of sites in $(Y_j)_{j\in J}$ to which X_i can successfully transmit, then we are interested in the set $\Psi^{(i)} - X_i$, see ◨ Fig. 6.1. This set contains the relevant information about the connectivity of X_i, in particular enough information to answer the above questions. It appears natural to center this set at the transmitter, i.e., we see it from the transmitter, and aim for an application of Theorem 6.7.7 to quantities of the form

$$L_n(A) = \frac{1}{|W_n|} \sum_{i\in I:\, X_i\in W_n} \mathbb{1}\{\Psi^{(i)} - X_i \in A\}, \qquad A \subset \mathbb{S}(\mathbb{R}^d). \tag{6.8.1}$$

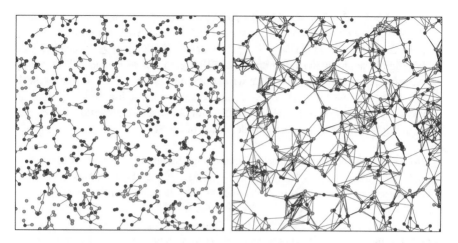

◻ Fig. 6.1 Realizations of a network of transmitters (blue) and receivers (green) with edges (black) indicating the possibility that a transmitter can connect to a receiver based on sufficiently large SINR. The two realizations use different intensities for the underlying PPPs and different SINR-thresholds

Instead of restricting to $X_i \in W_n$ in the sum, one could also require that the entire set $\Psi^{(i)}$ be contained in the box W_n or all the points that are needed to determine $\Psi^{(i)} - X_i$, but we do not want to bother about that as this will be asymptotically equivalent under the assumptions that we will pose, see Remark 6.5.3 and Exercise 6.5.4.

It is clear that the above questions can be formulated in terms of suitable choices of the set A. Indeed, for example, for the first question one would pick the set $A = \{\phi : \phi \neq \emptyset\}$, and for the second one $A = \{\phi : \#\phi \geq N\}$ respectively $A = \{\phi : \#\phi = N\}$. But many more choices of A are possible. Hence, L_n contains a lot of information and is worth studying.

Now $L_n(A)$ has the appearance of the empirical mark field, see Remark 6.7.8, for the random marked point process $(X_i)_{i\in I}$ where the mark of X_i is given by the set $\Psi^{(i)} - X_i$. Hence, the mark space $\mathcal{M}$ is here equal to $\mathbb{S}(\mathbb{R}^d)$, and the underlying stationary marked point process to which we try to apply Theorem 6.7.7 is $\Phi = ((X_i, \Psi^{(i)} - X_i))_{i\in I}$. Let us give some examples of how to define a successful transmission.

1. *Boolean model.* For some connectivity threshold $R \in (0, \infty)$, we call a transmission from X_i to Y_j successful if $|X_i - Y_j| \leq R$.

2. *Boolean model with random powers.* Assume that each X_i has an individual reach $R_i \in (0, \infty)$, i.e., X_i transmits to Y_j successfully if $|X_i - Y_j| \leq R_i$. Here it is natural to assume that the marks R_i are random and i.i.d.

3. *k-hop Boolean model.* We draw a bond between any two different points in $\{X_i : i \in I\} \cup \{Y_j : j \in J\}$ if their distance does not exceed $R \in (0, \infty)$, and we call a point X_i k-hop connected (to $(Y_j)_{j\in J}$) if there is a path in the corresponding Gilbert graph from X_i to $\{Y_j : j \in J\}$ with at most k hops, of which only the last one leaves $\{X_i : i \in I\}$. In this case, we call the transmission emitted from X_i successful. See Example 6.6.7 and ▶ Sect. 4.7 for more results on such a model.

4. *directed k-hop SINR-model.* Using a path-loss function ℓ as in ▶ Sect. 5.1 and an SINR-threshold $\tau \in (0, \infty)$, we draw a directed bond from any point X_i to any point $Z \in \{X_i : i \in I\} \cup \{Y_j : j \in J\}$ if $\text{SINR}(X_i, Z, (X_j)_{j \in J}) \geq \tau$, recall (5.2.2). We slightly modified the definition of the SINR graph introduced in ▶ Sect. 5.3 by dropping the symmetrization. Then we call the transmission from X_i successful if X_i is k-hop connected to $\{Y_j : j \in J\}$ in the same sense as for the k-hop Boolean model. It is an option to introduce and use individual power strengths $P_i \in (0, \infty)$ of the signal emitted from X_i in the definition of the SINR.

Note that, in all these examples, the underlying random marked point process $\Phi = ((X_i, \Psi^{(i)} - X_i))_{i \in I}$, or $\Phi = \big((X_i, (P_i, \Psi^{(i)} - X_i))\big)_{i \in I}$ if individual powers are used, is not an independently marked one, as the marks do depend on each other heavily, and in the k-hop variants variants with $k \geq 2$ and in the SINR-based ones, they depend also on other points of $(X_i)_{i \in I}$. However, if the path-loss function ℓ is assumed to have a bounded support, then all these dependencies are local, and even if not, then they can be approached with local cut-off versions, under natural assumptions on ℓ. Note that if the dependencies are local, then the mark space $\mathbb{S}(\mathbb{R}^d)$ may be taken substantially smaller.

Exercise 6.8.1
Prove that, in each of the above definitions of transmission success, the process $\Phi = ((X_i, \Psi^{(i)} - X_i))_{i \in I}$ is a stationary marked point process. Find natural conditions on the underlying random point processes $(X_i)_{i \in I}$ and $(Y_j)_{j \in J}$ and on ℓ that ensure its ergodicity. *Hint:* One may be happy with assuming the two processes as independent stationary PPPs. As it concerns ℓ, assuming it is bounded, decreasing and compactly supported, suffices. If one is more ambitious, then one imposes decay properties for ℓ and approximates with cut-off versions.   ◊

We now assume that the random point process Φ, distributed according to $\mathbb{Q}$, is amenable to an application of the ergodic theorem in Theorem 6.7.7. In order to find the limit of $L_n(A)$, we have to apply the theorem to the test function $G_A(m, \phi) = \mathbb{1}\{m \in A\}$, since $L_n(A) = \langle G_A, \mathcal{R}_{W_n}^{(s),o} \rangle$. Then we obtain that

$$\lim_{n \to \infty} L_n(A) = \lim_{n \to \infty} \langle G_A, \mathcal{R}_{W_n}^{(s),o} \rangle = \mathbb{Q}^o(A \times \mathbb{S}(\mathbb{R}^d \times \mathcal{M})). \tag{6.8.2}$$

Let us note that in this application, the mark $\Psi^{(o)}$ of the origin, which is distributed according to $\mathbb{Q}^o(\mathcal{M} \times \mathbb{S}(\mathbb{R}^d \times \mathcal{M}))^{-1} \mathbb{Q}^o(\cdot \times \mathbb{S}(\mathbb{R}^d \times \mathcal{M}))$, can be intuitively seen as the 'typical' set of points to which a typical transmitter can connect to, where the receiver process is independent of the transmitter process. One of the easiest cases is where $(X_i)_{i \in I}$ and $(Y_j)_{j \in J}$ are two independent stationary PPPs, but even here an evaluation of the limit is not an easy task for many sets A. Note that the ergodic limit depends in various manners on the parameters of the model, like k and R, the distribution of the power variables, and the details of ℓ. It offers a rich area of open research questions, for example about its behavior as $k \to \infty$, coupled with the intensity parameter of the process $(Y_j)_{j \in J}$ tending to zero, a question that was examined in ▶ Sect. 4.7 in a special case.

The empirical measure L_n defined in (6.8.1) will be further analyzed in ▶ Sect. 7.8 in view of frustration probabilities, i.e., in view of questions about the decay of probabilities of rare events. There we will also comment on the handling of boundary terms, an issue that we neglect in this section.

Note that if one would like to study the percentage of receivers Y_j in the window W_n that can successfully be reached by a given number of transmitters, then an analog approach can be used in which essentially the roles of transmitters and receivers have to be interchanged.

Let us finally mention an alternative ansatz for the description of the successful transmitters. For this, we conceive the superposition of transmitter and receiver processes as the underlying random point process, not only the transmitter process. This allows us to attack the above questions in the same situation, again with the help of Theorem 6.7.7, using a much simpler mark space on the cost of more complicated test functions G_A.

Let us give some details. As in Remark 6.6.6, we consider our reference process, the superposition $\{Z_i : i \in K\} = \{X_i : i \in I\} \cup \{Y_j : j \in J\}$ with marks 't' and 'r'. Observe that, even though this mark space is very simple, this random marked point process may have a complicated mark distribution that might even destroy ergodicity. But if we, e.g., assume that $(X_i)_{i \in I}$ and $(Y_j)_{j \in J}$ are independent stationary PPPs, then it is clear that we may apply Theorem 6.7.7 to $\Phi = ((Z_i, M_i))_{i \in K}$. If we decide to use individual power strengths P_i for the transmitters, then we add independently such a mark to each point of the process that is marked 't'. Now we define for each of the above notions of a success of a transmission observables $G : \mathcal{M} \times \mathbb{S}(\mathbb{R}^d \times \mathcal{M}) \to \mathbb{R}$ that register sets of successful transmitters. For $\phi = ((z_i, m_i))_{i \in I}$, and for $A \subset \mathbb{S}(\mathbb{R}^d)$, we let

$$G_A(m, \phi) = \mathbb{1}\{m = \mathrm{t}\} \mathbb{1}\{\{z_i : m_i = \mathrm{r}, \, o \text{ transmits successfully to } z_i\} \in A\},$$

where o is the additional point at the origin with mark 't', i.e., a transmitter. The set $\{z_i : m_i = \mathrm{r}, \, o \text{ transmits successfully to } z_i\}$ is picked out of the receivers, i.e., the points with mark 'r', but it depends on the entire point cloud ϕ. Now note that $\langle G_A, \mathcal{R}_{W_n}^{(\mathrm{s}),o}(\Phi) \rangle$ is in distribution equal to $L_n(A)$ defined in (6.8.1), i.e., we have achieved precisely what we were interested in. In case of the SINR model from above, if the path-loss function ℓ has a compact support, then G_A is a local function, otherwise approximation techniques are likely to be applicable. Now we can apply Theorem 6.7.7, for example, if $(X_i)_{i \in I}$ and $(Y_j)_{j \in J}$ are independent stationary PPPs with respective intensities λ and μ. Let $\mathbb{Q}$ be the distribution of $\Phi = ((Z_i, M_i))_{i \in K}$, then with the help of Theorem 6.7.7,

$$\lim_{n \to \infty} L_n(A) = \lim_{n \to \infty} \langle G_A, \mathcal{R}_{W_n}^{(\mathrm{s}),o} \rangle = \mathbb{Q}^o(\{\mathrm{t}\} \times A),$$

which provides an alternative expression for the right-hand side of (6.8.2). Again, a deeper analysis of the limit in dependence of the parameters is an interesting open research project, for example the consideration of the limit as $k \uparrow \infty$, coupled with $\mu \downarrow 0$, as we considered in ▶ Sect. 4.7 for the k-hop connection without interference.

Events of Bad Quality of Service: Large Deviations

For measuring the quality of a random telecommunication system, it is certainly important to quantify its expected performance, i.e., the quality of service in a normal situation. This is what we did in ▶ Chap. 6 in certain limit regimes. However, it appears equally important to know also something about *extreme situations*, i.e., about random occurrences of very rare events in such a limiting setting. In particular, we are interested in rare events of a bad service, like an event of the form in which only a small percentage of devices actually are connected, rare events of a particularly good service are less interesting. For such bad events, it is desirable to know (1) good upper bounds for the probability of this to happen, and (2) the characteristics of the situation that typically lead to this event. Good answers to these questions will provide the basis for a reliability analysis for the quality of the system, yield measures for its robustness against randomly occurring bad influences, and gives ansatzes how to design countermeasures.

In this chapter, we will present a mathematical theory that gives good answers to (1) and (2) on the base of asymptotic probability theory. Moreover, especially for (2), such theoretical investigations make important contributions, since they eludicate us from dealing with an enormous amount of data or simulations that would be necessary to gain a statistically reliable assertion about that rare event.

Mathematical tools for deriving such information are provided by the *theory of (the probabilities of) large deviations*. This theory is designed for the analysis of asymptotic random situations in which a parameter diverges that drives the unlikeliness of the event. If the theory of large deviations is applicable, it turns out that the probability under consideration decays even exponentially fast in this parameter, and the exponential rate is characterized in terms of a variational formula, which is amenable for a deeper investigation. The term "large deviations" expresses that the events under consideration are events of a large deviation of a random quantity X_n from its limiting value as a parameter n tends to infinity. But it is often used in much greater generality, describing just the decay of the probabilities of events of the form $\{X_n \in A\}$. Roughly speaking, large deviation theory provides tools for expressing the exponential rate $\lim_{n\to\infty} n^{-1} \log \mathbb{P}(X_n \in A)$, in terms of a variational formula that expresses the most likely way of X_n to lie in A.

© Springer Nature Switzerland AG 2020
B. Jahnel, W. König, *Probabilistic Methods in Telecommunications*, Compact Textbooks in Mathematics,
https://doi.org/10.1007/978-3-030-36090-0_7

In the context of telecommunication, we mainly think here of the two approximative situations that we encountered in ▶ Chap. 6: the *high-density limit*, where many devices are present in a bounded area, see ▶ Sect. 6.2, and the *ergodic limit* or the *thermodynamic limit*, where many devices are located in a large area with constant spatial density, see ▶ Sects. 6.4 and 6.6. However, we will discuss a number of different, relevant situations, for example excessively large interferences or small interference thresholds.

In this chapter, we explain what large deviations theory is and what it can achieve, see for example [DemZei10, DeuStr89] for general accounts and [RasSep15] for an introduction to the theory with statistical-physics flavor. In ▶ Sect. 7.1 we explain a typical example, in ▶ Sect. 7.2 we give a crash course on the theory, in ▶ Sect. 7.3 we discuss the high-density situation, with applications in ▶ Sect. 7.4 to highly connected devices, in ▶ Sect. 7.5 to frustration events coming from a lack of connectivity due to too much interference, and in ▶ Sect. 7.6 to large values of interferences. Furthermore, in ▶ Sect. 7.7 we explain the thermodynamic-limit situation with an application to connectivity in ▶ Sect. 7.8.

7.1   Introductory Example

The prototypical example is a random walk $S_n = X_1 + \cdots + X_n$, the partial sum of independent and identically distributed (i.i.d.) real random variables X_i. For definiteness, assume that the X_i have expectation equal to zero and additionally very strong integrability properties, more precisely, we assume that they have finite exponential moments of all orders, i.e., $\mathbb{E}[\exp(\alpha X_1)] < \infty$ for all $\alpha \in \mathbb{R}$. Then, according to the weak law of large numbers, S_n/n converges to zero in probability, i.e., the probability of the event $\{|S_n/n| \geq x\}$ converges towards zero as $n \to \infty$ for any $x > 0$. Such an event is the prototypical example of an *event of a large deviation*: the event that the quantity S_n/n deviates from its limit by an amount of at least x. In basic probability theory, its probability is upper estimated in terms of the Chebyshev inequality, the Markov inequality for the map $x \mapsto x^2$, under the sole assumption that the second moment is finite, i.e., $\mathbb{E}[X_1^2] < \infty$. The outcome is that it decays (at least) like $1/n$. Now we will find a much better upper bound, first for the upwards deviation $\{S_n \geq xn\}$.

With the help of the Markov inequality for the function $x \mapsto e^{xa}$ and the independence of the X_i, one derives the following upper bound for any $a > 0$ and $n \in \mathbb{N}$,

$$\mathbb{P}(S_n \geq xn) \leq e^{-axn}\mathbb{E}[e^{aS_n}] = e^{-axn}\mathbb{E}\Big[\prod_{i=1}^{n} e^{aX_i}\Big] = \Big(e^{-ax}\mathbb{E}[e^{aX_1}]\Big)^n. \qquad (7.1.1)$$

Note that we use that all the positive exponential moments of X_1 are finite. The technique that we demonstrated in (7.1.1) is sometimes called the *exponential Chebyshev inequality.* Optimizing over a, we derive an exponentially decaying upper bound, which we may summarize as

$$\limsup_{n\to\infty} n^{-1} \log \mathbb{P}(S_n \geq xn) \leq -I(x), \qquad x \in (0, \infty), \tag{7.1.2}$$

where the *rate function* I is given as $I(x) = \sup_{a>0}(ax - \log \mathbb{E}[e^{aX_1}])$. A bit of analytic work, see Exercise 7.1.4, reveals that $I(x)$ is really positive for $x > 0$. This already is an important fact, since exponential decay is much stronger than polynomial decay.

Remarkably, we can even derive the complementary lower bound, which shows that the decay is precisely exponential and identifies the exact decay rate, which is indeed equal to $I(x)$. In fact, somewhat deeper techniques, see Remark 7.1.6, show that in (7.1.2) also the opposite inequality holds and that a version for negative x holds as well. This may be loosely summarized by saying that

$$\mathbb{P}(S_n \approx xn) \approx e^{-nI(x)}, \qquad n \to \infty, \tag{7.1.3}$$

for any $x \in \mathbb{R}$, with rate function

$$I(x) = \sup_{a\in\mathbb{R}}(ax - \log \mathbb{E}[e^{aX_1}]), \qquad x \in \mathbb{R}, \tag{7.1.4}$$

which is the *Legendre transform* of the map $a \mapsto \log \mathbb{E}[e^{aX_1}]$. One says that $(S_n/n)_{n\in\mathbb{N}}$ *satisfies a large deviations principle (LDP)* with rate function I. This statement is also true if X_1 is not centered.

Exercise 7.1.1 (Fundamental Examples of Rate Functions)

1. Consider the simplest case where X_1 takes the values ± 1 with equal probability $p = 1/2$. Show that

$$I(x) = \begin{cases} 2^{-1}\big((1 + x)\log(1 + x) + (1 - x)\log(1 - x)\big) & \text{for } x \in [-1, 1], \\ \infty & \text{for } x \in \mathbb{R} \setminus [-1, 1], \end{cases}$$

where $0 \log 0 = 0$. In particular, note that $x \mapsto I(x)$ has a unique minimum in zero. Find $I(x)$ for general $p \in [0, 1]$.

2. Show that $I(x) = x^2/2\sigma^2$, for X_1 a normal random variable with expectation 0 and variance σ^2.

3. Show that $I(x) = \alpha x - 1 - \log(\alpha x)$ for $x > 0$ otherwise $I(x) = \infty$, for X_1 exponential with parameter α.

4. Show that $I(x) = \alpha - x + x\log(x/\alpha)$ for $x > 0$ otherwise $I(x) = \infty$, for X_1 Poisson with parameter α. ◇

Exercise 7.1.2

Let Φ be a homogeneous PPP, and put $W_n = [-n/2, n/2]^d$ for any $n \in (0, \infty)$, and we recall that $\Phi(W)$ is the number of Poisson points in $W \subset \mathbb{R}^d$. Use Exercise 7.1.1 to show that

$$\lim_{n \to \infty} n^{-d} \log \mathbb{P}(\Phi(W_n \setminus W_{n-b}) \geq \varepsilon n^d) = -\infty, \qquad b, \varepsilon > 0.$$

This shows that the probability of having of order n^d points in a boundary layer of the observation window decays super-exponentially fast on the volume scale. $\Diamond$

Exercise 7.1.3

Find an example such that $n^{-1} \log \mathbb{P}(S_n/n = x)$ has no limit for any $x \in \mathbb{R}$ as n tends to infinity. $\Diamond$

Exercise 7.1.4 (Properties of I)

Let $X = X_1$ be a random variable with all exponential moments finite and expectation $\mu \in \mathbb{R}$. Show that the Legendre transform I of the logarithm of the moment generating function, see (7.1.4), has the following properties: I is strictly convex, infinitely differentiable and nonnegative in $(\mathrm{essinf}\, X_1, \mathrm{esssup}\, X_1)$ and equal to $+\infty$ outside $[\mathrm{essinf}\, X_1, \mathrm{esssup}\, X_1]$. Furthermore, it attains its unique minimum at $x^* = \mu$ with value $I(x^*) = 0$. The maximum in (7.1.4) for $x = x^*$ is attained in the unique $a = a^*$ that satisfies $x^* = \mathbb{E}[X e^{a^* X}]/\mathbb{E}[e^{a^* X}]$.
$\Diamond$

From Exercise 7.1.4, one can deduce the law of large numbers that S_n/n converges in distribution to a^*, but we will do that in Remark 7.2.4 in broader generality.

It is important to note that the formula (7.1.4) for the exponential rate is explicit and amenable to further analysis, it contains, beyond a law of large numbers, useful and characteristic information about the way in which the large-deviations event is typically realized. The theories of convex functions and variational calculus are helpful here.

Exercise 7.1.5 (Best Realization of the Rare Event)

In the setting of Exercise 7.1.4, pick $x \in (\mu, \mathrm{esssup}\, X_1)$ and show that, conditioned on the rare event $\{S_n \geq xn\}$, the random walk obeys a law of large numbers towards the value $\mathrm{argmin}\{I(y) \colon y \geq x\}$. In other words, show that $x = \mathrm{argmin}\{I(z) \colon z \geq x\}$ and that $\lim_{n \uparrow \infty} \mathbb{P}(|S_n/n - x| > \varepsilon \mid S_n \geq xn) = 0$ for any $\varepsilon > 0$. $\Diamond$

Remark 7.1.6 (Details on the Lower Bound) The standard method to prove the lower bound in (7.1.3) is via the *Cramér transformation*, which is an exponentially transformed probability distribution of the step variables X_i that drives the random walk to adopt the deviation event as its typical behavior. This transform $\widehat{\mathbb{P}}_a$ has a parameter $a \in \mathbb{R}$ and is defined by using the Radon–Nikodym density $Z_a^{-1} e^{aX_1}$ with respect to the distribution of X_1, where $Z_a = \mathbb{E}[e^{aX_1}]$ is the normalizing constant. Hence, for any measurable set $A \subset \mathbb{R}$,

$$\widehat{\mathbb{P}}_a(X_1 \in A) = Z_a^{-1} \mathbb{E}\big[e^{aX_1} \mathbb{1}\{X_1 \in A\}\big]. \tag{7.1.5}$$

Now we fix $x \in \mathbb{R}$ and derive the lower bound in (7.1.3). It is elementary to calculate the expectation of X_1 under $\widehat{\mathbb{P}}_a$ explicitly in terms of the moment generating function $\varphi_a(x) = \widehat{\mathbb{E}}_a[e^{xX_1}]$ and to pick a in such a way that $\widehat{\mathbb{E}}_a(X_1) = x$. Indeed, it turns out that a is characterized as the unique number that realizes the maximum over y in (7.1.4). Now we rewrite the probability on the left in terms of this transformed measure,

$$\mathbb{P}(S_n \approx xn) = Z_a^n \widehat{\mathbb{E}}_a\big[e^{-aS_n} \mathbb{1}\{S_n \approx xn\}\big] \approx Z_a^n e^{-axn} \widehat{\mathbb{P}}_a(S_n \approx xn).$$

For the last term, if the event $\{S_n \approx xn\}$ is defined in a suitable way, e.g., as $\{|S_n/n - x| \leq x_n\}$ for some $x_n \to 0$ (not too quickly), one can apply well-known asymptotics in the vein of the law of large numbers, to see that its exponential rate is zero. Hence, the exponential rate of $\mathbb{P}(S_n \approx xn)$ is shown to be equal to $-(ax - \log Z_a)$. Using the above-mentioned characterization of a, we see that this is equal to $-I(x)$, and the proof of (7.1.3) is finished. ◇

Exercise 7.1.7
What is the Cramér transform of the Bernoulli distribution? ◇

Remark 7.1.8 (Cramér Transform and Gibbs Measures) Changes of measures of the type of a Cramér transform and generalizations play important roles in various subfields of probability theory and statistics. Indeed, the measure $\widehat{\mathbb{P}}_a$ with density $Z_a^{-1} e^{aX_1}$ (the Cramér transform in Remark 7.1.6) can be seen as a simple example in the very large class of so-called *Gibbs measures*, one of the central objects in statistical physics. In this context the density $Z_a^{-1} e^{aX_1}$ is often referred to as the *Boltzmann weight*. In general, Gibbs measures are constructed via Boltzmann weights of the form

$$Z_{n,\beta}^{-1} e^{-\beta \mathcal{H}(X_1,\dots,X_n)},$$

where the *Hamiltonian* $\mathcal{H}$ is a function interpreted as an energy of the configuration $(X_1, \dots, X_n)$, and $\beta \in (0, \infty)$ the inverse temperature. A particularly interesting set of models are the so-called *mean-field models*, where $\mathcal{H}(X_1, \dots, X_n) = nF(S_n/n)$ for some continuous function F of the sum $S_n = X_1 + \cdots + X_n$. For example, the choice $F(x) = x^2$ gives the famous *Curie–Weiss model*. It is one of the corner stones of the theory of Gibbs measures that, in the limit as n tends to infinity, phase transitions such as spontaneous magnetization can be observed, using large deviations theory. See [Geo11, Hän12, Der19] for some general expositions to Gibbs measures in various different contexts. ◇

Remark 7.1.9 (Importance Sampling) The idea of the proof of the lower bound as described in Remark 7.1.6 via the Cramér transform also hints towards a theoretical solution for the problem of how to effectively sample the large deviation event. For this, note that large deviation events share the property that they are typically exponentially unlikely, and hence very hard to sample. Indeed, we need $\exp(\mathcal{O}(n))$ i.i.d. samples to observe the large deviation event and we want to establish statistics based on many samples of that event, which requires a large number of repetitions. This is computationally close to infeasible. On the other hand, sampling from the measure $\widehat{\mathbb{P}}_a$ given by the Cramér transform for an optimal choice of a

makes the unlikely event likely and thus much easier to observe, since we can decrease the sample size by a factor of $\exp(\mathcal{O}(n))$. In this way, large deviation principles can be used to device a so-called *importance sampling algorithm*, where samples are drawn according to an exponentially changed measure based on the rate function. In our example the new measure would have the density with respect to the distribution of X_1 given by

$$Z_a^{-1} e^{aX_1} \quad \text{where} \quad ax - \log Z_a = I(x).$$

More details on importance sampling can, for example, be found in [AsmGly07, KTB13]. ◊

7.2   Principles of Large Deviations

In this section, we give the fundamental elements and assertions of the theory of large deviations in a reasonably general framework.

We want to formulate a large deviations principle, i.e., the fact that the exponential decay of probabilities involving random variables X_n is governed by some rate function I. This is properly done in terms of the weak convergence of the set function $n^{-1} \log \mathbb{P}(X_n \in \cdot)$ towards the set function $-\inf\{I(x) : x \in \cdot\}$. Recalling the Portmanteau theorem on characterizations of weak convergences of probability measures, we obtain the idea to coin a definition in terms of upper bounds for closed sets and a lower bound for open sets.

Definition 7.2.1 (Large Deviations Principle)

Let $\mathcal{X}$ be a topological space and $(X_n)_{n \in \mathbb{N}}$ be a sequence of $\mathcal{X}$-valued random variables. Furthermore, let $I : \mathcal{X} \to [0, \infty]$ be a lower semicontinuous function. We say that $(X_n)_{n \in \mathbb{N}}$ (or equivalently, the sequence of its distributions) *satisfies a large deviations principle* (LDP) with *speed* n and *rate function* I, if, for any open set $G \subset \mathcal{X}$ and for any closed set $F \subset \mathcal{X}$,

$$\liminf_{n \to \infty} n^{-1} \log \mathbb{P}(X_n \in G) \geq - \inf_{x \in G} I(x), \tag{7.2.1}$$

$$\limsup_{n \to \infty} n^{-1} \log \mathbb{P}(X_n \in F) \leq - \inf_{x \in F} I(x). \tag{7.2.2}$$

A function $f : \mathcal{X} \to [-\infty, \infty]$ is called lower semicontinuous if for all $x^* \in \mathcal{X}$ we have that $\limsup_{x \to x^*} f(x) \leq f(x^*)$. Hence, topology plays a central role in the definition of an LDP.

Exercise 7.2.2 (Uniqueness of the Rate Function)
Show that the rate function is unique.   ◊

Remark 7.2.3 (Large Deviations for General Sets) Certainly, if $(X_n)_{n \in \mathbb{N}}$ satisfies an LDP with rate function I, we would like to have that $\mathbb{P}(X_n \in A) = \exp\left(- n(\inf_A I + o(1)) \right)$

as $n \to \infty$ for many events A, but this is not true in general and has to be verified on a case-by-case basis. Certainly, under additional assumptions on A and on I, this is indeed true, e.g., if the infima of I on the open kernel A° and on the closure $\overline{A}$ of A are identical. A theory that is based on the validity of $\mathbb{P}(X_n \in A) = \exp\left(-n(\inf_A I + o(1))\right)$ for any measurable A would be too restrictive to be helpful or interesting. Rather one sticks to a topology-oriented approach and a sense of exponential convergence that reminds on the weak topology of probability measures. $\diamond$

Remark 7.2.4 (LDP $\implies$ Convergence) If $(X_n)_{n\in\mathbb{N}}$ satisfies an LDP on $\mathcal{X}$ with rate function I, under additional assumptions, one can derive a law of large numbers from the LDP, more precisely the weak convergence of X_n towards the minimizer of I. Some obvious, and mostly used, sufficient condition, in the case that $\mathcal{X}$ is a metric state space, is that a unique minimizer x^* of I exists and that the minimum is stable in the sense that $\inf\{I(x) : |x - x^*| > \varepsilon\} > I(x^*)$ for any $\varepsilon > 0$. Then it directly follows from the upper bound that X_n converges towards x^* with exponential decay of the probability to deviate from x^* by some positive amount. Deriving the convergence from the LDP is sometimes easier than attacking the convergence of X_n more directly, we present an example for this in ▶ Sect. 7.4. In the case that I has more than one minimizer, but still some stability, a kind of convergence towards the set of minimizers follows in the same way. Hence, a discussion about the minimizers of the rate function I is absolutely vital for a deeper understanding of the asymptotic behavior of X_n. $\diamond$

Remark 7.2.5 (LDP $\implies$ Analysis of Frustration Events) One of our main motivations to discuss large deviations theory in connection with telecommunication is its value for the analysis of rare events of a bad system quality, i.e., events of the form $\{X_n \in A\}$ on which, e.g., there is a bad connectivity or too much of interference. We call such an event a *frustration event*. Certainly, also 'good' events can be analyzed, but they are not so important from our viewpoint. Hence, we are faced with questions like (1) how large is the probability of this event, and (2) how does X_n typically behave on this event? Question (1) was briefly discussed in Remark 7.2.3 on a general level. Question (2) is the question about the behavior of X_n under the conditional measure $\mathbb{P}(\cdot \mid X_n \in A)$. Having understood the point in Remark 7.2.3, the natural reaction is the guess that X_n should, under this measure, converge towards the minimizer of I on the set A, if it is unique, and a similar statement if not. This is true in many situations, but again the correct formulation and its justification need additional assumptions on A and on I and a proof on a case-by-case basis. The natural line of argument would be to prove first that I has on A just one minimizer x_A and to justify the statement

$$\mathbb{P}(X_n \in B_\varepsilon(x_A)^c \mid X_n \in A) = \frac{\mathbb{P}(X_n \in A \setminus B_\varepsilon(x_A))}{\mathbb{P}(X_n \in A)}$$

$$= \exp\left(-n\Big[\inf_{A\setminus B_\varepsilon(x_A)} I - \inf_A I\Big]\right) e^{o(n)}, \qquad n \to \infty,$$

and to find criteria under which the difference of the two infima is strictly positive.

An analysis of x_A will then give a deeper answer to Question (2). However, its dependence on the rate function I, the explicitness of its description, and its regularity with respect to A, may render this a major challenge that cannot in every case be mastered in a satisfactory way.   ◊

Exercise 7.2.6 (Lower Semicontinuity and Convexity of the Rate Function)
In the setting of Exercise 7.1.4, show that $x \mapsto I(x)$ is strictly convex and lower semicontinuous.   ◊

Example 7.2.7 (Cramér's Theorem)
The introductory example of ▶ Sect. 7.1 is called *Cramér's theorem*. It states that, if a sequence of i.i.d. centered real-valued random variables $(X_i)_{i\in\mathbb{N}}$ is given, having all exponential moments finite, then the sequence of the empirical means $(X_1 + \cdots + X_n)/n = S_n/n$ satisfies an LDP with speed n and rate function I given by (7.1.4), the Legendre transform of the map $y \mapsto \log \mathbb{E}[\exp(yX_1)]$. The function I is convex and non-negative and possesses the expected value of X_1 as its only zero. The proof comes in several steps:
1. The upper bound for sets $F = [x, \infty)$ with $x > 0$ are proved in (7.1.1).
2. Sets of the form $(-\infty, -x]$ are handled in the same way.
3. The proof of the corresponding lower bound is outlined in Remark 7.1.6.
4. General open, and closed, sets are handled by using that I is strictly increasing in $[0, \infty)$ and strictly decreasing in $(-\infty, 0]$.   ◊

Exercise 7.2.8
Prove Cramér's theorem as outlined in Example 7.2.7.   ◊

A famous and useful extension of Cramér's theorem is an LDP for the *empirical measure* of many i.i.d. random variables.

Theorem 7.2.9 (Sanov's Theorem)
Let $\mathcal{X}$ be a Polish space and $(X_n)_{n\in\mathbb{N}}$ an i.i.d. sequence of $\mathcal{X}$-valued random variables with marginal distribution μ. Then their normalized empirical measure $L_n = n^{-1}\sum_{i=1}^{n} \delta_{X_i}$ satisfies an LDP on the set of probability measures on $\mathcal{X}$ (equipped with the weak topology) with rate function equal to

$$I(\nu) = H(\nu|\mu) = \int f(x) \log f(x)\mu(\mathrm{d}x), \tag{7.2.3}$$

if the density $\mathrm{d}\nu/\mathrm{d}\mu = f$ exists, and $H(\nu|\mu) = \infty$ otherwise.

A *Polish space* is a complete separable metric space. The term $H(\nu|\mu)$ is called the *Kullback–Leibler divergence* or *relative entropy* of ν with respect to the reference measure μ.

Sketch of proof
Instead of a formal proof, let us demonstrate why the statement is true in the simpler setting of a finite state space $\mathcal{X}$. In this case, we can apply a direct calculation. Indeed, for any

probability measure v on $\mathcal{X}$ with $v(x) > 0$ for any $x \in \mathcal{X}$, we have (ignoring that $v(x)$ does not have to be in $n^{-1}\mathbb{N}_0$)

$$\mathbb{P}(L_n \approx v) = \mathbb{P}\left(\forall x \in \mathcal{X} \colon \#\{i \in \{1,\ldots,n\} \colon X_i = x\} = nv(x)\right)$$

$$= \binom{n}{(nv(x))_{x \in \mathcal{X}}} \prod_{x \in \mathcal{X}} \mu(x)^{nv(x)} = \frac{n!}{\prod_{x \in \mathcal{X}}(nv(x))!} \prod_{x \in \mathcal{X}} \mu(x)^{nv(x)}, \qquad (7.2.4)$$

since the collection of the $nL_n(x)$, $x \in \mathcal{X}$, is multinomially distributed. Now Stirling's formula in the form $n! = (\frac{n}{e})^n e^{o(n)}$ yields that

$$\mathbb{P}(L_n \approx v) = \frac{(n/e)^n}{\prod_{x \in \mathcal{X}}(nv(x)/e)^{nv(x)}} e^{o(n)} \prod_{x \in \mathcal{X}} \mu(x)^{nv(x)} = e^{-nI(v)} e^{o(n)}, \qquad (7.2.5)$$

where we introduced

$$I(v) = \sum_{x \in \mathcal{X}} v(x) \log \frac{v(x)}{\mu(x)} = H(v|\mu), \qquad (7.2.6)$$

the entropy of v with respect to μ, the rate function. □

Remark 7.2.10 (The Relative Entropy in Statistical Physics) The relative entropy $H(v|\mu)$ plays an important role not only in context of large deviations, but also, for example, in information theory, informatics, statistical physics and beyond. The term 'entropy' is almost universally present in the physics of thermodynamics. Many related but different concepts share the same name. The notion of relative entropy as we present it here appears naturally in the theory of large deviations, which is intimately related to statistical physics. To a large extent, these fields are connected since they share the same notion of relative entropy.

Let us employ the way of thinking of statistical physics to better understand relative entropy. Inspecting the proof sketch of Sanov's theorem, we see that the relative entropy emerges from two parts. First, the number of *microstates*, i.e., the number of sets $(X_1, \ldots, X_n)$, giving rise to the *macrostate* v, which results on an exponential scale in an *entropy* term of the form $H(v) = -\sum_{x \in \mathcal{X}} v(x) \log v(x)$. Second, the probability to see such a microstate, resulting in an *energy* term of the form $\sum_{x \in \mathcal{X}} v(x) \log \mu(x)$, the minus sign is a convention to make $H(v)$ non-negative. Note that $H(\mu|\mu) = 0$, or in other words, if the target macrostate is such that the probability to see an appropriate microstate matches the number of appropriate microstates, then there is no discrepancy between the two parts and no convergence to zero. Otherwise, the relative entropy measures the difference between the two competing forces, the entropy and energy of the system. ◊

Remark 7.2.11 (The Relative Entropy in Information Theory) As mentioned in Remark 7.2.10, relative entropies appear in various incarnations in seemingly unrelated subdomains of probability theory and statistics. In view of our general subject telecommunication, let us briefly elaborate on how and why relative entropies emerge in information theory and how this connects to the statistical physics viewpoint.

First note that the entropy, $H(v)$, as defined in Remark 7.2.10, reflects the number of possible microstates associated to a given macrostate v. This can be interpreted as the amount of uncertainty or disorder in the statistical ensemble. For example, if $v = \delta_x$, i.e., the macrostate is a delta measure in $x \in \mathcal{X}$, then $H(v) = 0$ and there is no uncertainty about how this state is produced from the microstate. On the other hand, the entropy is maximal if v is the uniform distribution.

Now making the connection to information theory, we reinterpret $\mathcal{X}$ as an alphabet and v as a probability distribution on its symbols, for example modeling the frequency of letters $x \in \mathcal{X}$ in english text. From this perspective, the entropy can serve as a measure for the uncertainty in the letter distribution. One way to make this precise is via *Shannon's source coding theorem* [Sha48]. For this we are interested in ways to encode each symbol $x \in \mathcal{X}$ as a binary sequence $c(x) = \{0, 1\}^{l(x)}$ of (variable) length $l(x)$ in such a way that the original alphabet can be recovered without loss, i.e., $c = (c(x))_{x \in \mathcal{X}}$ is a prefix code where no whole codeword is the beginning of another codeword. Additionally we are interested in a code such that the expected length $L_v(c) = \sum_{x \in \mathcal{X}} v(x)l(x)$ is small. Then, the statement of the theorem is that

$$L_v(c) \geq H_2(v),$$

and there exists a code such that

$$L_v(c) \leq 1 + H_2(v).$$

Here H_2 refers to the entropy where the natural log is replaced by the logarithm with base 2. In words, the theorem states that the entropy provides theoretical bounds on optimal coding-schemes and thus captures the expected cost to transmit data through the communication channel. For more details see for example [Gra11, Mac03].   ◇

The relative entropy is not a metric, for example, because it is not symmetric in its arguments. However, the following exercises should help to shed more light on properties of H and to convince us that it shares important properties of a metric between measures.

Exercise 7.2.12
Prove that the relative entropy is non-negative using Jensen's inequality for the convex function $x \mapsto x \log x - x + 1$ on $[0, \infty)$. Further prove that $H(v_1 + v_2 | \mu) \geq H(v_1 | \mu) + H(v_2 | \mu)$ and that $v \mapsto H(v | \mu)$ is convex and lower-semicontinuous.   ◇

Exercise 7.2.13 (Properties of the Entropy)
In each of the examples presented in Exercise 7.1.1, explicitly verify that the rate function can be expressed as a relative entropy. More precisely, for each x find a probability measure v_x such that $I(x)$ is its relative entropy with respect to the distribution of the step variable X_1. *Hint:* Consider the Cramér transform of Remark 7.1.6.   ◇

Exercise 7.2.14

Prove the following bound of the relative entropy with respect to the total variational distance

$$H(\nu|\mu) \geq 2^{-1}\|\nu - \mu\|^2,$$

where $\|\nu - \mu\| = \sup\{|\nu(A) - \mu(A)|: A \subset \mathcal{X} \text{ measurable}\}$. Convince yourself that an inverse bound cannot hold in general. ◊

Here is a rather important and helpful representation of the relative entropy as the Legendre transform of the logarithmic moment-generating function.

Lemma 7.2.15 (The Relative Entropy as a Legendre Transform) *For any two probability measures ν, μ on a Polish space $\mathcal{X}$,*

$$\begin{aligned} H(\nu|\mu) &= \sup_{f \in \mathcal{C}_b} [\langle f, \nu \rangle - \log\langle \exp(f), \mu \rangle] \\ &= \sup_{f \in \mathcal{B}_b} [\langle f, \nu \rangle - \log\langle \exp(f), \mu \rangle], \end{aligned} \tag{7.2.7}$$

where $\mathcal{C}_b$ and $\mathcal{B}_b$ denote the sets of bounded and continuous, bounded and measurable, respectively, functions on $\mathcal{X}$, and we write $\langle f, \nu \rangle$ for the integral of a function f with respect to a measure ν.

Indeed, for nice measures ν and μ, the supremum is attained in $f = \log d\nu/d\mu$, otherwise one has to approximate, which renders the proof of the lemma pretty technical, see e.g., [DeuStr89, Lemma 3.2.13]. This lemma can be used for performing proofs of certain LDPs in the same way as we described in Remark 7.1.6, i.e., with the help of a Cramér transform. It also has the nice consequence that it makes it possible to see Sanov's theorem as an abstract version of Cramér's theorem.

Remark 7.2.16 (Sanov $\Longleftrightarrow$ Cramér) Sanov's theorem can be seen as an abstract version of Cramér's theorem, since also the probability measures $(\delta_{X_i})_{i \in \mathbb{N}}$ form an i.i.d. sequence, but on the space of probability measures on $\mathcal{X}$. This brings us to the idea that the two formulas for the rate functions, the Legendre transform in (7.1.4) and the explicit formula in (7.2.3), might be identical to each other, if (7.1.4) is adapted to a probability measure δ_{X_1} instead of a random variable X_1. This is indeed true, and the right adaptation is in terms of a bilinear pairing between the state space (here the space of probability measures on $\mathcal{X}$, seen as a subset of the vector space of signed measures) and its topological dual (here the set of bounded continuous functions). In the setting of Sanov's theorem, the pairing map is $\mathcal{C}_b \times \mathcal{M}_1 \ni (f, \nu) \mapsto \langle f, \nu \rangle = \int f(x) \nu(dx)$, while in the setting of Cramér's theorem for real random variables, this pairing is just the map $\mathbb{R} \times \mathbb{R} \ni (x, y) \mapsto xy$. The analogue that arises is indeed equal to the first line of (7.2.7) in Remark 7.2.15, which states that the entropy is the Legendre transform of the logarithm of the exponential moments of the test integrals against continuous bounded functions. With this understanding, we have identified Sanov's theorem as an abstract version of Cramér's theorem. ◊

Exercise 7.2.17

Verify that the empirical mean S_n is the expected value of the empirical measure L_n.   ◊

One useful tool says that continuous maps turn LDPs into new LDPs by taking a continuous image.

> **Theorem 7.2.18 (Contraction Principle)**
>
> If $(X_n)_{n\in\mathbb{N}}$ satisfies an LDP with rate function I in the topological state space $\mathcal{X}$, and if $F\colon \mathcal{X} \to \mathcal{Y}$ is a continuous map into another topological space, then also $(F(X_n))_{n\in\mathbb{N}}$ satisfies an LDP, and the rate function $J\colon \mathcal{Y} \to [0,\infty]$ is given by
>
> $$J(y) = \inf\{I(x)\colon x \in \mathcal{X},\, F(x) = y\}, \qquad y \in \mathcal{Y}.$$

Exercise 7.2.19

Prove the contraction principle. *Hint:* This does not require fancy arguments. It may be helpful to observe first that it suffices to check continuity of F in every $x \in \mathcal{X}$ with $I(x) < \infty$.   ◊

The contraction principle derives LDPs for 'simpler' functionals from LDPs for 'more comprehensive' quantities, and it gives an explicit formula for the rate function. We will frequently make use of this idea, since in some relevant situations it is easier to obtain an LDP for a quantity that contains a lot of detailed information and to contract to the LDP for a less cumbersome version, than proving the latter directly.

LDPs for averages $n^{-1}\sum_{i=1}^{n} f(X_i)$, as appearing in Cramér's theorem, are sometimes called *level-1 LDPs*, while LDPs for empirical measures of the form $n^{-1}\sum_{i=1}^{n}\delta_{X_i}$, as appearing in Sanov's theorem, are sometimes referred to as *level-2 LDPs*. Then the evaluation maps contract a level-2 LDP to a level-1 LDP. Indeed, if f is a test function such that the map $\nu \mapsto \langle f,\nu\rangle$ is continuous (e.g., if f is continuous and bounded, if the usual weak topology is considered) then, if $(\nu_n)_n$ satisfies an LDP, then $(\langle f,\nu_n\rangle)_n$ satisfies one. In this way, the contraction principle basically derives Cramér's theorem from Sanov's theorem. However, the distinction according to the levels is only formal, as level 2 can also be seen as a special case of level 1 on a more abstract space, as we saw in Remark 7.2.16.

One of the cornerstones of the theory tells how to evaluate the exponential rate of expectations of an exponential function of variables that satisfy an LDP. We call a rate function $I\colon \mathcal{X} \to [0,\infty]$ *good* if its level sets $\{x \in \mathcal{X}\colon I(x) \le \alpha\}$ are compact.

> **Theorem 7.2.20 (Varadhan's Lemma)**
>
> If $(X_n)_{n\in\mathbb{N}}$ satisfies an LDP with good rate function I in the topological state space $\mathcal{X}$, and if $f\colon \mathcal{X} \to \mathbb{R}$ is continuous and bounded, then

(Continued)

Theorem 7.2.20 (continued)

$$\lim_{n\to\infty} n^{-1} \log \mathbb{E}\big[e^{nf(X_n)}\big] = \sup_{x\in\mathcal{X}} \big(f(x) - I(x)\big).$$

This is a substantial extension of the well-known *Laplace principle* that says that $\int_0^1 \exp(nf(x))\,dx$ behaves to first order like $\exp(n\max_{[0,1]} f)$ if $f\colon [0, 1] \to \mathbb{R}$ is continuous. The minimizers of $f - I$ may be a source of deeper information about the realizations of X_n that give the main contribution to the expectation of $\exp(nf(X_n))$. In many applications, this requires a formidable analytic work and does not always lead to satisfactory results, since neither f nor I are continuous nor bounded.

Another important tool is a technique to derive an LDP from the validity of Varadhan's lemma. More precisely, if all limiting logarithmic asymptotics of exponential moments of suitable test functions exist and have a certain regularity, then the corresponding LDP is true. We call a sequence $(X_n)_{n\in\mathbb{N}}$ of random variables with values in a locally topological space $\mathcal{X}$ *exponentially tight* if for any $C \in (0, \infty)$ there is a compact set $K \subset \mathcal{X}$ such that

$$\limsup_{n\to\infty} n^{-1} \log \mathbb{P}(X_n \in K^c) \le -C. \tag{7.2.8}$$

This criterion allows to extend the upper bound in (7.2.2) from compact sets F to closed sets F. In the following we assume that the state space $\mathcal{X}$ is a Banach space and that the dual space $\mathcal{X}^*$ of $\mathcal{X}$, the set of all continuous bounded functions $\mathcal{X} \to \infty$, induces the topology on $\mathcal{X}$ via the pairing $(F, x) \mapsto \langle F, x \rangle$, i.e., that $x_n \to x$ in $\mathcal{X}$ if and only if $\langle F, x_n \rangle \to \langle F, x \rangle$ for any $F \in \mathcal{X}^*$.

Theorem 7.2.21 (Gärtner–Ellis Theorem)
Let $\mathcal{X}$ be a Banach space, and let $(X_n)_{n\in\mathbb{N}}$ be an exponentially tight sequence of $\mathcal{X}$-valued random variables such that the limit

$$\Lambda(F) = \lim_{n\to\infty} n^{-1} \log \mathbb{E}\big[e^{nF(X_n)}\big] \in \mathbb{R}, \qquad F \in \mathcal{X}^*, \tag{7.2.9}$$

exists and induces a lower semicontinuous and Gâteau-differentiable function $\Lambda\colon \mathcal{X}^* \to \mathbb{R}$. Then $(X_n)_{n\in\mathbb{N}}$ satisfies an LDP on $\mathcal{X}$ with rate function Λ^*, given by

$$\Lambda^*(x) = \sup_{F\in\mathcal{X}^*} [\langle F, x \rangle - \Lambda(F)], \qquad x \in \mathcal{X}. \tag{7.2.10}$$

A function $\Lambda\colon \mathcal{X}^* \to \mathbb{R}$ is called Gâteau-differentiable if for any $F, g \in \mathcal{X}^*$, the limit $\lim_{t\to 0} t^{-1}[\Lambda(F+tg) - \Lambda(F)]$ exists. This theorem is a far-reaching generalization of Cramér's theorem and has a proof that follows the same route, with substantially more technicalities.

Remark 7.2.22 (Fenchel–Legendre Transform) The function Λ^* is the *Fenchel–Legendre transform* of Λ. The duality principle states that, for general lower semicontinuous convex functions Λ, the Fenchel–Legendre transform of Λ^* is equal to Λ, i.e., $(\Lambda^*)^* = \Lambda$, at least if $\mathcal{X}$ is a locally convex Hausdorff topological vector space. $\diamond$

Remark 7.2.23 (Fenchel–Legendre Transform and Entropy) All the main tools and results that we are presenting in this chapter are best suitable for averages of many random identically distributed objects (variables, Delta measures, ...) that are either independent or close to independent. This is obvious in Cramér's and Sanov's theorems, and it is implicit in the Gärtner–Ellis theorem. A consequence is that, in all these results, the rate function can be written as a Fenchel–Legendre transform or as an entropy, which is roughly the same, as we saw in Remark 7.2.15, and is therefore convex. $\diamond$

Remark 7.2.24 (The Role of the Gâteau Differentiability) The upper bound in (7.2.2) can be derived without problems just by assuming the existence of the limit in (7.2.10), but the proof of the lower bound in (7.2.1) needs the additional assumptions, in particular the Gâteau differentiability, and more work. The sense of the Gâteau differentiability is not to ensure more regularity, but it is characteristic for a situation in which the rate function is convex. The theorem does not apply, for example, for a measure on $[0, 1]$ of the form $\mu_n(\mathrm{d}x) = Z_n^{-1} \exp(-nI(x))\,\mathrm{d}x$ with a normalizing constant Z_n and a smooth function $I: [0, 1] \to [0, \infty)$ that is not convex. Indeed, $(\mu_n)_{n \in \mathbb{N}}$ does satisfy an LDP with rate function I, but the function Λ is not differentiable in those points F such that the infimum of $\langle F, x \rangle - \Lambda(F)$ is attained in *two* points x_1 and x_2, but not in between. $\diamond$

Now we briefly discuss another important theoretical tool, which implies an LDP for 'infinitely long' objects from the LDPs for each cut-off version. The abstract setting for that is the setting of projective limits, see [DemZei10, Section 4.6]. Here we assume that, on the index set J (countable or uncountable), there is a partial order inequality such that for any $i, j \in J$ there is a $k \in J$ such that $i \leq k$ and $j \leq k$. Then a family $((\mathcal{Y}_j)_{j \in J}, (\Pi_{i,j})_{i \leq j})$ is called a *projective system* if every $\mathcal{Y}_j$ is a Hausdorff topological space and $\Pi_{i,j}: \mathcal{Y}_j \to \mathcal{Y}_i$ is continuous such that $\Pi_{i,j} \circ \Pi_{j,k} = \Pi_{i,k}$ for $i \leq j \leq k$. Then the *projective limit* $\mathcal{X}$ is defined as the space of those $x = (y_j)_{j \in J}$ such that $y_j \in \mathcal{Y}_j$ for any $j \in J$ and $y_i = \Pi_{i,j}(y_j)$ for any $i \leq j$. We equip $\mathcal{X}$ with the product topology from the topological product of the spaces $\mathcal{Y}_j$. By $\Pi_j: \mathcal{X} \to \mathcal{Y}_j$ we denote the canonical projection.

Standard examples are the set of functions $[0, T] \to \mathbb{R}^d$ with $J = \{(t_1, \ldots, t_n): n \in \mathbb{N}, 0 \leq t_1 < \cdots < t_n \leq T\}$ and $\mathcal{Y}_{t_1,\ldots,t_n} = \mathbb{R}^n$ on one hand, and the sequence space $\mathcal{X} = \mathbb{R}^{\mathbb{N}_0}$ with $J = \mathbb{N}_0$ and $\mathcal{Y}_j = \mathbb{R}^j$. Later the set of point processes on $\mathbb{R}^d$, particularly on bounded boxes, will be relevant for us.

Now the main result about LDPs involving projective limits is the following.

Theorem 7.2.25 (LDP for Projective Limits; Dawson–Gärtner)
Let $\mathcal{X}$ be the projective limit of the projective system $((\mathcal{Y}_j)_{j \in J}, (\Pi_{i,j})_{i \leq j})$. Let $(\mu_n)_{n \in \mathbb{N}}$ be a family of probability measures on $\mathcal{X}$ such that, for any $j \in J$, the family $(\mu_n \circ \Pi_j^{-1})_{n \in \mathbb{N}}$ satisfies, for $n \to \infty$, an LDP on $\mathcal{Y}_j$ with good rate function I_j. Then $(\mu_n)_{n \in \mathbb{N}}$ satisfies an LDP as $n \to \infty$ on $\mathcal{X}$ with rate function

$$I(x) = \sup_{j \in J} I_j(\Pi_j(x)), \qquad x \in \mathcal{X}. \tag{7.2.11}$$

One of the standard applications, see also ▶ Sect. 7.7, is to derive an LDP for empirical measures $n^{-1} \sum_{i=1}^{n} \delta_{\theta^i((X_n)_{n \in \mathbb{N}})}$ of infinitely long sequences of random variables, where θ^i is the i-fold shift operator, from the LDPs for the empirical measures $n^{-1} \sum_{i=1}^{n} \delta_{(X_i, X_{i+1}, \ldots, X_{i+k-1})}$ of the shifts of k-strings for any $k \in \mathbb{N}$. Another natural application (in fact, the d-dimensional, continuous version) is to derive an LDP for the empirical field $|W|^{-1} \int_W \delta_{\theta_x(\Phi)} \, dx$ of a PPP Φ in $\mathbb{R}^d$ from the LDPs for their projections to some window.

Let us briefly mention one technical tool for deriving LDPs in cases where the textbook results are not directly applicable for some reason of technical nature, but not in principle. This is often the case if a crucial functional is not bounded or not continuous, or if a crucial random variable is not bounded. Often such obstacles are amenable to an approximation with quantities that do satisfy an LDP and are close to the quantities under interest. However, this approximation must be 'on an exponential level'. Let us specify which kind of closeness is sufficient, see [DemZei10, Section 4.2.2].

Definition 7.2.26 (Exponentially-good Approximations)

Let $(\mathcal{X}, d)$ be a metric space.

(i) Two sequences of random variables $(X_n)_{n \in \mathbb{N}}$ and $(Y_n)_{n \in \mathbb{N}}$ with values in $\mathcal{X}$ are called *exponentially equivalent* to each other, if they can be defined on one probability space in such a way that, for any $n \in \mathbb{N}$ and $\delta \in (0, \infty)$, the event $\{d(X_n, Y_n) > \delta\}$ is measurable and satisfies

$$\limsup_{n \to \infty} n^{-1} \log \mathbb{P}(d(X_n, Y_n) > \delta) = -\infty.$$

(ii) A family of sequences $(X_n^{(r)})_{n \in \mathbb{N}}, r \in \mathbb{N}$, of $\mathcal{X}$-valued random variables is called an *exponentially-good approximation* of a sequence $(X_n)_{n \in \mathbb{N}}$ if all the variables can be defined on one probability space such that all the events $\{d(X_n, X_n^{(r)}) > \delta\}$ are measurable for any $n, r, \in \mathbb{N}$ and $\delta \in (0, \infty)$ and

$$\lim_{r \to \infty} \limsup_{n \to \infty} n^{-1} \log \mathbb{P}(d(X_n, X_n^{(r)}) > \delta) = -\infty.$$

Certainly, this concept can also be formulated for other scales, and for $\varepsilon \to 0$ instead of $n \to \infty$.

Example 7.2.27 (Cutting Interference)

One standard way of defining helpful exponentially good approximations is by cutting unbounded objects down to finite size. We indicate this for the question of having a extremely large interference, leaving the details as an exercise. This example will be developed further in ▶ Sect. 7.6.

Let $\Phi = (X_i)_{i \in I}$ be a standard PPP in $\mathbb{R}^d$ and ℓ a path-loss function as in ▶ Sect. 5.1. We consider the interference at the origin o, $I(o) = \sum_{i \in I} \ell(\|X_i\|)$, and we want to study the large deviations of $(\varepsilon I(o))_{\varepsilon \in (0,\infty)}$ as ε tends to zero. This means that we look, for example, at the probability of the event $\{I(o) \geq a/\varepsilon\}$ for some $a \in (0, \infty)$ in this limit. We see that this event should come mainly from a large number of points X_i close to the origin, i.e., from a high-density setting. If the support of ℓ is unbounded, in particular if the integral of $\mathbb{R}^d \ni x \mapsto \ell(\|x\|)$ is not finite, then one has to take into account the entire sum on i, and it may be that technical problems arise when separating parts close to the origin and far away from it. The technicalities may be formulated and carried out in terms of exponentially good approximations by cutting. For example, one can replace ℓ by the cut-off version $\ell_r = \ell \mathbb{1}_{[0,r]}$ and define $I_r(o)$ as the interference measured in terms of ℓ_r instead of ℓ. Then $\varepsilon I_r(o)$ is a good candidate for an exponentially-good approximation. Equivalently, one could replace the summation over all i by the summation over those i such that $\|X_i\| \leq r$. If ℓ is unbounded close to zero, a further cutting at small values may be necessary.

An exercise could consist of working out the details, in particular determining the large-deviation rate and the assumptions on ℓ under which the concept carries through. $\diamond$

Example 7.2.28 (Empirical Pair Measures)

Empirical measures with different boundary conditions are often exponentially equivalent. An example that is simple to verify is the following. Let $(X_n)_{n \in \mathbb{N}}$ be a Markov chain on some finite set, then the two empirical pair measures $L_n = n^{-1} \sum_{i=1}^{n} \delta_{(X_{i-1}, X_i)}$ and $L_n^{(s)} = n^{-1}(\delta_{(X_n, X_1)} + \sum_{i=2}^{n} \delta_{(X_{i-1}, X_i)})$ are exponentially equivalent. Note that $L_n^{(s)}$ is just the cyclic variant of L_n, where X_0 is replaced by X_n. $\diamond$

Exercise 7.2.29 (Empirical Fields)

Here is another example in the same vein as Example 7.2.28, which we will use in Exercise 7.7.5: Consider the empirical field $\mathcal{R}_{W_n}(\Phi)$ defined in (6.4.7) and its stationary version $\mathcal{R}_{W_n}^{(s)}(\Phi)$ defined in (6.7.2), for Φ a standard PPP on $\mathbb{R}^d$ and $W_n = [-n, n]^d$. Prove that they are exponentially equivalent. *Hint:* The assertion and the proof of Exercise 6.7.2 are helpful. $\diamond$

Certainly, one expects that each member of a pair of exponentially equivalent sequences should satisfy the same LDP, if any, and that the LDPs for exponentially-good approximations should be close to each other. This latter one is true, but it needs a careful formulation and a technical proof, which we omit.

> **Theorem 7.2.30 (LDP and Exponentially-good Approximations)**
> *Let $(X_n^{(r)})_{n\in\mathbb{N}}$ be an exponentially good approximation as $r \to \infty$ of $(X_n)_{n\in\mathbb{N}}$, and assume that, for any r, $(X_n^{(r)})_{n\in\mathbb{N}}$ satisfies an LDP with rate function $I_r: \mathcal{X} \to [0,\infty]$. Then $(X_n)_{n\in\mathbb{N}}$ satisfies a weak LDP (i.e., the upper bound is true a priori only for compact sets) with rate function*
>
> $$I(x) = \sup_{\delta>0} \liminf_{r\to\infty} \inf_{y\in B_\delta(x)} I_r(y), \qquad x \in \mathcal{X}. \tag{7.2.12}$$

If the level sets of I are compact and $\inf_F I \le \lim_{r\to\infty} \inf_F I_r$ for any closed set $F \subset \mathcal{X}$, then $(X_n)_{n\in\mathbb{N}}$ satisfies even the full LDP, as one might prove as an exercise. The convergence of I_r towards I in (7.2.12) is called *Gamma convergence*; it is specially adapted to the convergence of minima.

Exercise 7.2.31 (LDP and Exponential Equivalence)

Prove that exponentially equivalent sequences satisfy the same LDP, if any.   ◊

Exercise 7.2.32

Formulate and prove that the assertion of Sanov's theorem remains true if up to $n\varepsilon_n$ of the i.i.d. random variables $X_1, \ldots, X_n$ are replaced by some other random variables, where $\varepsilon_n \to 0$.   ◊

7.3   LDP in the High-Density Setting

Now let us consider the high-density limit for the simple model that we encountered in Example 6.2.1. Let a bounded communication area D and a PPP $(X_i)_{i\in I_\lambda}$ in D with absolutely continuous intensity measure $\lambda\mu$ be given. Here λ is a positive parameter, which will be sent to infinity. According to Lemma 6.2.2, the *normalized empirical measure* $L_\lambda = \lambda^{-1} \sum_{i\in I_\lambda} \delta_{X_i}$ converges towards the intensity measure. That is, $L_\lambda \Longrightarrow \mu$ as $\lambda \to \infty$ in the τ-topology, i.e., when testing against measurable and bounded functions. In particular, we have this convergence also in the weak topology (i.e., when restricting the set of test functions to the continuous bounded ones), and in this topology we want to consider now an LDP.

Lemma 7.3.1 (LDP for L_λ) *Let $(X_i)_{i\in I_\lambda}$ be a PPP in a compact set $D \subset \mathbb{R}^d$ with intensity measure $\lambda\mu$, where $\lambda \in (0,\infty)$ and μ is an absolutely continuous measure on D. Then the normalized empirical measure $L_\lambda = \lambda^{-1} \sum_{i\in I_\lambda} \delta_{X_i}$ satisfies, as $\lambda \to \infty$, the LDP on the set of measures on D (equipped with the weak topology) with rate function given by*

$$I(m) = H(m|\mu) = \int f(x) \log f(x)\, \mu(\mathrm{d}x) - m(D) + \mu(D), \tag{7.3.1}$$

if the density $\mathrm{d}m/\mathrm{d}\mu = f$ exists, and $H(m|\mu) = \infty$ otherwise.

The term $H(m|\mu)$ is an extension of the Kullback–Leibler divergence appearing in (7.2.3) in Sanov's theorem to positive measures.

Proof

Instead of a proof, let us give an argument why this LDP should be true, at least in the weak topology. Pick some measure m on D with density with respect to μ. We want to heuristically evaluate the probability of the event $\{L_\lambda \approx m\}$. We approximate this by picking a decomposition of D into many small measurable sets $A_1, \ldots, A_n$ with positive Lebesgue measure. Then we find, using the independence of $L_\lambda(A_1), \ldots, L_\lambda(A_n)$ and the Poisson nature of $(X_i)_{i \in I_\lambda}$,

$$
\mathbb{P}(L_\lambda \approx m) \approx \mathbb{P}\big(L_\lambda(A_k) \approx m(A_k)\, \forall k = 1, \ldots, n\big)
$$

$$
= \prod_{k=1}^{n} \mathbb{P}\Big(\sum_{i \in I_\lambda} \delta_{X_i}(A_k) \approx \lambda m(A_k)\Big)
$$

$$
= \prod_{k=1}^{n} \mathbb{P}\big(\#\{i \in I_\lambda : X_i \in A_k\} \approx \lambda m(A_k)\big)
$$

$$
= \prod_{k=1}^{n} \Big[\frac{(\lambda \mu(A_k))^{\lambda m(A_k)}}{(\lambda m(A_k))!}\, e^{-\lambda \mu(A_k)}\Big].
$$

Now use the rough form of Stirling's formula, $n! = (n/e)^n e^{o(n)}$, to see that

$$
\mathbb{P}(L_\lambda \approx m) \approx \prod_{k=1}^{n} \Big[\Big(\frac{\lambda \mu(A_k) e}{\lambda m(A_k)}\Big)^{\lambda m(A_k)} e^{-\lambda \mu(A_k)}\Big]
$$

$$
\approx \exp\Big(-\lambda \sum_{k=1}^{n}\Big[m(A_k) \log \frac{m(A_k)}{\mu(A_k)} + \mu(A_k) - m(A_k)\Big]\Big)
$$

$$
= \exp\big(-\lambda H(m^{(n)}|\mu^{(n)})\big),
$$

where $m^{(n)}$ and $\mu^{(n)}$ are the coarsening projections of m and μ on the decomposition $(A_1, \ldots, A_n)$ of D. It is an exercise to see that their entropy converges towards the entropy of m and μ in the limit $n \to \infty$ of zero fineness of the decomposition. □

Remark 7.3.2 (Connection with Sanov's Theorem) The high-density LDP can also be derived from Sanov's theorem, see Theorem 7.2.9, using the construction of the PPP in Lemma 2.2.6 via first fixing the number of points in D, which is Poisson-distributed with parameter $\lambda\mu(D)$, and then using that the points are i.i.d. with distribution $\mu/\mu(D)$. However, observe that the empirical measure L_λ of the Poisson points is not normalized to a probability measure, unlike the one for i.i.d. random variables. ◊

Exercise 7.3.3

Show that the relative entropy as defined in (7.3.1) is convex and lower semicontinuous. ◊

Exercise 7.3.4

Formulate an extension of Lemma 7.3.1 for an independently marked PPP $((X_i, M_i))_{i \in I_\lambda}$ and outline an argument analogously to our sketch for Lemma 7.3.1. (If you are more ambitious, then you do that for a general marked PPP.) ◇

Example 7.3.5

Let us briefly mention an LDP for the cloud of connectable receivers for a signal stemming from the origin, in the limit of a small threshold for the SINR. This is taken from [HJKP16] and we give here only rough indications.

Let $\Phi = (X_i)_{i \in I}$ and $\Psi = (Y_j)_{j \in J}$ denote two independent homogeneous PPPs, where the X_i are the transmitters and the Y_j the receivers. We fix a bounded and monotonously decreasing path-loss function ℓ satisfying $\ell(r) \sim r^{-\alpha}$ as $r \to \infty$ with some $\alpha \in (d, \infty)$, and we measure the success of the transmission from X_i to a point of Ψ in terms of the signal-to-interference ratio, SINR, as in ▶ Sect. 5.2. We add a transmitter at the origin, o. Consider the set of connectable receivers of the origin,

$$
Y^\tau = \{Y_j \in \Psi : \mathrm{SINR}(o, Y_j, \Phi \cup \{o\}) > \tau\}
$$

$$
= \left\{ Y_j \in \Psi : \ell(|Y_j|) > \tau \left(N + \sum_{i \in I} \ell(|X_i - Y_j|) \right) \right\},
$$

where $\tau \in (0, \infty)$ is the threshold and $N \in (0, \infty)$ some noise constant. We would like to examine an LDP for a rescaling of this set in the limit $\tau \downarrow 0$. Indeed, in this limit, this set should become larger and larger, more precisely, it should have a radius of order $\tau^{-1/\alpha}$, since the denominator is always of order one, and the numerator in the SINR is not larger than order one for all the points Y_j that are in a centered ball of that distance. Let us conceive $Y^\tau(\tau^{-1/\alpha} \cdot)$ as a random measure on $\mathbb{R}^d$, i.e., $Y^\tau(\tau^{-1/\alpha} A) = \#\{Y_j \in \Psi : Y_j \in \tau^{-1/\alpha} A, \mathrm{SINR}(o, Y_j, \Phi \cup \{o\}) > \tau\}$. Then the right normalization of this measure is the volume of the ball, i.e., we are searching for an LDP for the measure $\tau^{d/\alpha} Y^\tau(\tau^{-1/\alpha} \cdot)$. Let us note at this place that such an LDP would also, as a corollary, give the exponential rate of the probability that the origin is disconnected.

Indeed, such an LDP has been derived in [HJKP16, Theorem 2], even under many more general assumptions (in the presence of additional features like random fading variables and random signal strengths). At this point, we can only give some rough indications about the structure of the LDP and its background.

We see already that a kind of high-density LDP as in Lemma 7.3.1 for the empirical measure of the Y_js in large compact subsets of $\mathbb{R}^d$ is involved. However, the description of its distribution depends on the additional random input, the process Φ, via the interference, and it depends on it on large space scales, more precisely at least in the centered ball with radius of order $\tau^{-1/\alpha}$. In order to handle this correctly, one has to employ an additional LDP for the locations of the transmitters X_i, in the spirit of the LDP that we will discuss and handle in Theorem 7.7.4. Additionally, an LDP for the intensities of the point processes involved has to be employed. ◇

7.4   Empirical Degree Measures in a Highly-Dense System

Here is an application of the LDP of ▶ Sect. 7.3 to a telecommunication-related question, giving also quite some space for future research. We want to discuss an important tool for measuring the local connectedness in a highly-dense telecommunication system in the light of large deviations theory. Here we restrict to the simplest setting, i.e., we neglect interference and long trajectories, but we only register the number of neighbors to which a direct communication is possible. Thereby, we are going to extend Example 6.2.4.

Consider the Boolean model for the PPP Φ on a compact subset D of $\mathbb{R}^d$ (see ▶ Sect. 4.1) with an absolutely continuous intensity measure $\lambda\mu$ and connectivity threshold R. Here we rely on the interpretation of the model as a random geometric graph, a Poisson–Gilbert graph, see ▶ Sect. 4.3. That is, we draw an edge between any two vertices of Φ if their distance is not larger than some fixed $R > 0$. Let us denote by $\deg(X_i)$ the degree of vertex $X_i \in \Phi$, that is, the number of direct neighbors of X_i.

In the high-density setting $\lambda \to \infty$, each node X_i has of the order λ neighbors. We are interested in the *empirical degree distribution*,

$$L_\lambda^{(\mathrm{deg})} = \lambda^{-1} \sum_{i \in I_\lambda} \delta_{\lambda^{-1} \deg(X_i)}, \tag{7.4.1}$$

which we already rescaled properly for considering the limit as $\lambda \to \infty$.

$L_\lambda^{(\mathrm{deg})}$ is a random measure on the set $\mathcal{M}([0, \infty))$ of measures on $[0, \infty)$. It registers the relative frequency of occurrences of a given distribution as the degree distribution of the Poisson–Gilbert graph given by the vertices X_i. It can be used to quantify unexpectedly large proportions of nodes with large degree. In the context of device-to-device networks, these nodes may serve as relays for their many neighbors and thus have a tendency to fail due to capacity constraints. Having too many such network components might lead to an unstable network performance. Thus it is desirable to understand how unlikely these events are and what typical configurations look like in these unlikely events. The base of this is the establishment of an LDP, which is done in the following lemma. As usual, $\mathcal{M}(D)$ and $\mathcal{M}([0, \infty))$ are both equipped with the weak topology.

Lemma 7.4.1   (LDP for the Degree Distribution) *In the above situation, $L_\lambda^{(\mathrm{deg})}$ satisfies, in the limit $\lambda \to \infty$, an LDP on the set of measures on $\mathcal{M}([0, \infty))$ with speed λ and rate function*

$$m \mapsto \inf\{H(\nu|\mu) \colon \nu \in \mathcal{M}(D), \ F(\nu) = m\},$$

where $F \colon \mathcal{M}(D) \to \mathcal{M}([0, \infty))$ is given by

$$F(\nu)(A) = \int \nu(\mathrm{d}y) \, \mathbb{1}\{\nu(B_R(y) \cap D) \in A\}, \qquad A \subset [0, \infty) \text{ mb.}$$

Proof

Basically, we use the contraction principle (Theorem 7.2.18) and the LDP for the empirical measure L_λ of Lemma 7.3.1, but some technicalities have to be overcome (which we leave as exercises). Indeed, observe that, on the event $\{L_\lambda \approx \nu\}$, the number of neighbors of X_i (i.e., $\deg(X_i)$) behaves like $\lambda \nu(B_R(X_i) \cap D)$. Hence, $L_\lambda^{(\text{deg})}$ is approximately equal to the empirical measure of the terms $\nu(B_R(X_i) \cap D)$, $i \in I_\lambda$, on this event. Therefore, we have $L_\lambda^{(\text{deg})} \approx F(\nu)$ on this event. This illustrates that the contraction principle should be applicable to the LDP for L_λ under the contracting map F, resulting in the claim. However, one has to justify the approximations above, and one has to prove that F is continuous at any absolutely-continuous $\nu \in \mathcal{M}(D)$.   □

Exercise 7.4.2

Fill in the details in the proof of Lemma 7.4.1.   ◇

As a corollary of Lemma 7.4.1, we easily obtain, again via the contraction principle, an LDP for the *average degree*

$$L_\lambda^{(\text{av deg})} = \lambda^{-2} \sum_{i \in I_\lambda} \deg(X_i),$$

which is the image of $L_\lambda^{(\text{deg})}$ under the map $\nu \mapsto \langle \text{id}, \nu \rangle$, where id denotes the identical map. Indeed, the rate function is given by

$$y \mapsto \inf\{H(\nu|\mu) : \nu \in \mathcal{M}(D), \ G(\nu) = y\}, \tag{7.4.2}$$

where $G(\nu) = \int \nu(\mathrm{d}x) \nu(B_R(x) \cap D)$. (Alternately, one can employ a similar proof as the one of Lemma 7.4.1 that we indicated above.) Also the average degree is an important characteristic for the connectedness of the system, and it is easier to handle than the empirical degree distribution.

Recalling Remark 7.2.4, the two LDPs tell us that we might have a convergence of the empirical degree distribution, of the average degree, towards the minimizer(s) of the respective rate functions with an exponential decay of the probabilities of a deviation, provided the minima are stable in the sense that we described in Remark 7.2.4. Since $L_\lambda \Longrightarrow \mu$ according to Lemma 6.2.2, we have, according to the continuous mapping theorem, that $L_\lambda^{(\text{deg})} = F(L_\lambda) \Longrightarrow F(\mu)$ and $L_\lambda^{(\text{av deg})} = G(L_\lambda) \Longrightarrow G(\mu)$; we already hinted at the latter in Example 6.2.4.

Recalling Remark 7.2.5, the above LDPs can also be used for an analysis of the probability and the entire situation where the system has a bad connectivity in the sense that $L_\lambda^{(\text{deg})} \approx \nu$ for some $\nu \in \mathcal{M}([0, \infty))$ different from $F(\mu)$ or that $L_\lambda^{(\text{av deg})} \approx y$ for some $y \in [0, \infty)$ different from $G(\mu)$. However, the kind of analytical questions that arise here are generally difficult and have not fully been understood, in particular for interesting generalization that we cannot discuss here.

It is crucial that the connectivity threshold R is kept fixed in the LDP above. In case of a rescaled threshold R_λ, the situation becomes far more complex and phenomena like homogenization and localization for the lower and upper part, of the LDP can be observed, see [SepYuk01, ChaHar14, HJT19].

Exercise 7.4.3
Assume that D is a centered disk and μ is isotropic. Show that the set of minimizers of the rate function (7.4.2) is also rotationally invariant, i.e., if ν is a minimizer, then also any rotation of ν is a minimizer. $\Diamond$

Exercise 7.4.4
Formulate and derive the LDP for the empirical location-degree distribution,

$$\frac{1}{\lambda} \sum_{i \in I_\lambda} \delta_{(X_i, \lambda^{-1} \deg(X_i))},$$

which registers also the place in D at which a particularly high connectivity is present. $\Diamond$

7.5   Connectivity to a Base Station in Dense Systems

Now let us demonstrate another example of assertions that can be deduced from the LDP of Lemma 7.3.1 in the spirit of Remark 7.2.5. This example is taken from [HJKP18]. We will analyze the frustration event of too many devices being unable to send their messages to a single base station, due to too much interference. We will distinguish the two cases of direct communication to the base station and of a multihop functionality, which we reduce to at most two hops for simplicity here.

Assume that a single base station is placed at the center o of the compact communication area D and that a PPP $\Phi = (X_i)_{i \in I_\lambda}$ of devices in D is given with absolutely continuous intensity measure $\lambda \mu$. The network is assumed to carry a relaying functionality, that is, messages do not have to be delivered to the base station directly but can also use one intermediate relaying step. We call a device frustrated if any possible message route from it is blocked due to too low SIR along the message trajectory. We are working with the SIR (signal-to-interference ratio) introduced in ▶ Sect. 5.2 (with zero noise, $N = 0$), more precisely with the total interference version discussed in Remark 5.3.1. However, since we are going to consider the high-density situation in which many devices are present and transmit messages, we decided to add a downscaling factor of $\gamma = 1/\lambda$ in front of the interference, in order to cope with the exploding amount of interference coming from an exploding number of devices. This is of course only a mathematical trick in order not to have to introduce time and to divide the message emissions on many time instances, as proposed in ▶ Sect. 5.4.

Hence, we can rewrite the SIR as a functional of the empirical measure $L_\lambda = \lambda^{-1} \sum_{i \in I_\lambda} \delta_{X_i}$ of the PPP of devices as $\text{SIR}(X_i, x, L_\lambda)$, where we denote

$$\text{SIR}(X_i, x, \nu) = \frac{\ell(|X_i - x|)}{\langle \ell(|\cdot -x|)], \nu \rangle}, \qquad \nu \in \mathcal{M}(D),$$

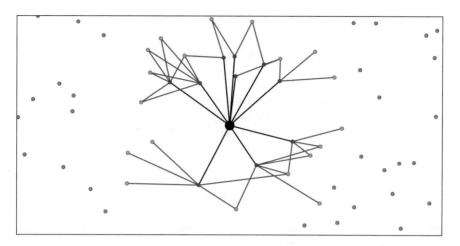

◻ Fig. 7.1 Realization of device configuration with blue devices directly linked to the origin, green devices indirectly connected and red devices disconnected

and recall the notation $\langle f, v \rangle$ for the integral of a function f with respect to v.

Now we want to allow each message from some X_i to the base station at the origin o to make one direct step or at most one relaying step into some relay X_j. We require that each of the two steps $X_i \rightarrow X_j$ and $X_j \rightarrow o$ has to satisfy the interference condition that the SIR is not smaller than a given threshold $\tau \in (0, \infty)$. Let us write this in terms of the minimum SIR in a trajectory $x \rightarrow y \rightarrow o$

$$D(x, y, o, v) = \min\{\mathrm{SIR}(x, y, v), \mathrm{SIR}(y, o, v)\}.$$

The maximum over the two trajectories $x \rightarrow o$ and $x \rightarrow y \rightarrow o$ for some y is then given by

$$R(x, o, v) = \max\left\{\mathrm{SIR}(x, o, v), \max_{y \in \Phi} D(x, y, o, v)\right\}.$$

We will render the transmission of a message from X_i to o successful if $R(X_i, o, L_\lambda) \geq \tau$ for a given threshold $\tau \in (0, \infty)$, i.e., if either the SIR of the direct or both SIRs of the indirect, two hops are larger than τ. In ◻ Fig. 7.1 we present a realization of the network indicating direct and indirect uplinks.

We are interested in the large deviation behavior of the empirical measure of frustrated devices, that is, of those whose message do not reach o. Their empirical measure may be written in terms of the function

$$\varphi_{v,\tau}(x) = \mathbb{1}\{R(x, o, v) < \tau\}, \qquad x \in D,$$

as the measure with density $\varphi_{L_\lambda,\tau}$ with respect to L_λ, that is,

$$\mathfrak{M}_{L_\lambda}(dx) = \varphi_{L_\lambda,\tau}(x)\, L_\lambda(dx) = \frac{1}{\lambda} \sum_{i \in I_\lambda} \delta_{X_i}(dx) \mathbb{1}\{R(x, o, L_\lambda) < \tau\}.$$

Note that L_λ appears here at two places: as the ground measure and as inducing interference; a general definition of the measure $\mathfrak{M}_\nu$ for arbitrary measures ν is obvious. Examples of events that are handled in [HJKP18] are of the form

$$\{\mathfrak{M}_{L_\lambda}(D) - \mathfrak{M}_\mu(D) > \varepsilon\}.$$

In words, this is the event that the proportion of disconnected devices is by $\varepsilon > 0$ larger than expected. This is a typical frustration event, and one would like to know what the most likely behavior of the cloud L_λ of devices is that realizes this event. Now using the LDP of Lemma 7.3.1, one naturally guesses that for $b > 0$,

$$\lim_{\lambda \uparrow \infty} \frac{1}{\lambda} \log \mathbb{P}(\mathfrak{M}_{L_\lambda}(D) > b) = -\inf\{H(\nu|\mu) : \nu \in \mathcal{M}(D), \mathfrak{M}_\nu(D) > b\}. \quad (7.5.1)$$

This is indeed true, but its proof in [HJKP18] requires a lot of technical work, since it is not possible to apply Lemma 7.3.1 directly. Indeed the event $\{\mathfrak{M}_{L_\lambda}(D) > b\}$ is not closed, and the map $\nu \mapsto \mathfrak{M}_\nu(D)$ is not continuous, and one has to carry out an approximating procedure.

Exercise 7.5.1

Verify that indeed the map $\nu \mapsto \mathfrak{M}_\nu(D)$ is discontinuous. *Hint:* Based on the construction, decreasing the number of devices in an area to zero might lead to a substantial number of devices becoming disconnected due to the fact that no relays are available.   ◊

Recalling that we would like to understand the frustration situation better, we need to analyze the variational formula on the right-hand side of (7.5.1) more closely. As we roughly explained in Remark 7.2.5, the (approximate) minimizers ν of $\inf\{H(\nu|\mu): \mathfrak{M}_\nu(D) > b\}$ describe the typical behavior of the system conditioned on the atypical event $\{\mathfrak{M}_{L_\lambda}(D) > b\}$, where we are most interested in the case where $b > \mathfrak{M}_\mu(D)$. Unfortunately, this problem is rather difficult, if not untractable, since the space $\mathcal{M}(D)$ is pretty large and unhandy, and it is notoriously cumbersome to work with entropies explicitly. It seems that there are only some few structural assertions easily derived, and any further-reaching property seems to require inconvenient techniques. For instance, in the special case that μ is the Lebesgue measure on a disc D, it can be shown that any minimizer is rotationally invariant if only a direct uplink is allowed, see Exercise 7.5.2.

However, the variational formula gives rise to simulation work that gives a good picture of the situation under consideration. For instance, a plot of the resulting radial density in the mentioned radial-symmetric case is given in ◘ Fig. 7.2 in the case of a disk with radius 5. It can be clearly seen that an event of too many disconnected devices

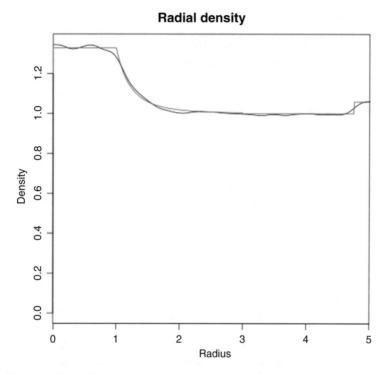

Fig. 7.2 Plot of the radial intensity as a function of the radius from [HJKP18, Fig. 2]. The black curve is based on simulations whereas the red curve is based on a corresponding analytic approximation

is typically achieved by putting slightly more devices at the cell boundary. Even more prominently, it is entropically beneficial to also put more devices next to the base station at the center. Those devices create a higher than expected interference, leading to more devices being disconnected at the cell boundary.

Exercise 7.5.2
Assume a rotation-invariant setting where D is a disc and μ is isotropic. Further consider a direct uplink setting where $R(x, o, \nu)$ is simply replaced by $SIR(x, o, \nu)$. Show that any minimizer of the rate function is also isotropic. *Hint:* In this setting, only the distance to the origin is relevant. Then, any rotation-invariant version of a given measure can only decrease the relative entropy. $\Diamond$

7.6 High Exceedances of Interference

In this section, we examine the extreme situation that the interference coming from many devices is enormously large. Here we mainly follow [GanTor08]. The results that we present do not have to do with the LDP in Lemma 7.3.1, but we feel that this lemma will be helpful for future, deeper investigation; see the comments at the end of this section.

Let $\Phi = (X_i)_{i \in J}$ be a homogeneous PPP in $\mathbb{R}^d$ with intensity $\lambda \in (0, \infty)$. We equip each X_i with a random power $P_i \in (0, \infty)$ of its message transmission and assume that the point process $((X_i, P_i))_{i \in J}$ is a marked PPP such that the powers are independent of the locations, i.e., an independently marked PPP, see ► Sect. 2.4. We want to study the interference at the origin,

$$I = I(o) = \sum_{i \in J} P_i \ell(|X_i|),$$

where $\ell \colon (0, \infty) \to [0, \infty)$ is a path-loss function of the type that we introduced in ► Sect. 5.1, e.g., one of the examples in 5.1.1. We are now interested in the asymptotics of the extreme event $\{I \geq s\}$ as $s \to \infty$ as well as in a description of the most likely configuration of the locations X_i and their powers P_i that lead to this event. In particular we ask:

1. What is the probability to experience such a high value of the interference?
2. Under what circumstances does the high interference originate from just one enormously strong signal, and from where is it transmitted?
3. If it comes from many signals, where are these many transmitters located?

Let us give some rough heuristics that explain possible answers. We need to collect all possible configurations that lead to the event $\{I \geq s\}$, determine the asymptotics of their probabilities and then pick the optimal one. First we consider the event that all the devices X_i that are close by also take part in the event (i.e., have a large power P_i) and that there are many; we want to neglect the path-loss function ℓ for a while, i.e., put it equal to one on a centered compact ball D and zero otherwise. We expect here that the number n of devices in D is very large and that practically all of them transmit a very strong signal of size $\approx p$. In order to meet the value s of the signal, we have to assume that $np \approx s$. The probability of having precisely n devices in D is equal to $\exp(-\lambda|D|)(\lambda|D|)^n/n!$, as this number is Poisson distributed with parameter equal to $\lambda|D|$. The asymptotics of this are dominated by the term $1/n!$, which is equal to $(e/n)^n \exp(o(n)) = \exp(-(1 + o(1))n \log n)$, according to the rough version of Stirling's formula. On the other hand, the probability that all these n devices transmit a signal of strength at least p is roughly equal to $\mathbb{P}(P > p)^n$, if P is one of the power variables. Hence, the probability of the event that n devices are present in D and transmit at strength p with $s = np$ is equal to

$$\exp\big\{-(1 + o(1))n \log\big(n/\mathbb{P}(P > s/n)\big)\big\}.$$

A natural ansatz is to require that $n \in \mathbb{N} \cap (0, s)$ is picked such that $n \log(n/\mathbb{P}(P > s/n))$ is minimal, asymptotically for $s \to \infty$. (We took s as an upper bound, since the powers should be large.) Under natural assumptions on the asymptotics of $x \mapsto \mathbb{P}(P > x)$, this will determine the choice of n and will give a preliminary, rough answer. In particular, this asymptotics is decisive for the answers to the above questions.

Next, let us consider power distributions of the type

$$-\log \mathbb{P}(P > x) \sim Cx^\gamma, \qquad x \to \infty, \text{ with parameters } C, \gamma \in (0, \infty). \qquad (7.6.1)$$

These are the distributions with upper tails of *Weibull* type, which include, for example the exponential distribution for $\gamma = 1$ and the normal distribution for $\gamma = 2$. Then, the asymptotic minimum of $n \mapsto n \log(n/\mathbb{P}(P > s/n))$ is characterized by

$$0 = \frac{d}{dn}\left[n \log n + Cs^\gamma n^{1-\gamma}\right], \qquad \text{i.e.,} \quad n^\gamma(1 + \log n) = (\gamma - 1)Cs^\gamma.$$

Now we see that we have to distinguish two cases with respect to the value of γ. First, if $\gamma \in (0, 1]$, then the right-hand side is non-positive, and the equation cannot be satisfied; the minimum is attained at $n = 1$, and the (single!) power is equal to s. Furthermore, if $\gamma \in (1, \infty)$, then the optimal n is asymptotically equal to $n_s = (C(\gamma - 1))^{1/\gamma} s/(\log s)^{1/\gamma}$, which lies in $(0, s)$ for sufficiently large s, and the corresponding powers are of order $\log s$. Then the corresponding scale of the negative exponential rate of $\mathbb{P}(I > s)$ is $\asymp n_s \log n_s \asymp s(\log s)^{1-1/\gamma}$.

The above heuristics is for $\ell(r) = 1$ only, but it already gives the right picture and conclusions. Let us summarize the main body of the result of [GanTor08].

Theorem 7.6.1 (Large Deviations for the Interference)

Let $\Phi = (X_i)_{i \in J}$ be a PPP in $\mathbb{R}^d$ with intensity λ, assume that $\ell(r) = \max\{R, r\}^{-\alpha}$ for some $R \in (0, \infty)$ and $\alpha \in (d, \infty)$, and assume that, given Φ, the family $(P_i)_{i \in J}$ is an i.i.d. family of random variables whose distribution is given by (7.6.1). Then the family $(s^{-1}I)_{s>1}$ satisfies the LDP as $s \to \infty$

- *if $\gamma \in (0, 1]$: with speed s^γ and rate function $[0, \infty) \ni x \mapsto CR^{\alpha\gamma}x^\gamma$,*
- *if $\gamma \in (1, \infty)$: with speed $s(\log s)^\eta$ and rate function $[0, \infty) \ni x \mapsto \gamma(\gamma - 1)^\eta C^{1/\gamma} R^\alpha x$, where $\eta = 1 - 1/\gamma$.*

Let us just drop the remark that the proof of this LDP in [GanTor08] implicitly uses the concept of exponentially good approximations, which we introduced at the end of ▶ Sect. 7.2. Further note that for $\gamma = 1$, the two cases coincide. The influence of the path-loss function $\ell(r) = \max\{R, r\}^{-\alpha}$ is not on the speed or the case distinction, but only on the constants in the rate function.

In order to see that Theorem 7.6.1 answers at least the first of our questions from above, just note that the negative large-s exponential rate of $\mathbb{P}(I > s) = \mathbb{P}(s^{-1}I \in (1, \infty))$ on the scale s^γ respectively $s(\log s)^\eta$ is not larger than the infimum of the rate function on the interval $(1, \infty)$ and not smaller than its infimum over $[1, \infty)$, and hence

$$-\log \mathbb{P}(I > s) \sim \begin{cases} s^\gamma CR^{\alpha\gamma}, & \text{if } \gamma \leq 1, \\ s(\log s)^\eta C^{1/\gamma} R^\alpha & \text{if } \gamma > 1. \end{cases}$$

The interpretation of Theorem 7.6.1 is that, for $\gamma > 1$ (when the power random variables possess all exponential moments), the large value of the interference comes from a superposition of many large signals from many transmitters, and the probability term that describes their large amount is only the term $1/n!$ appearing in the Poisson probability for having n transmitters. Interestingly, the density parameter λ does not appear on the leading scale. On the other hand, in the case $\gamma < 1$ (when the power variables are heavy-tailed in the sense that all their exponential moments are infinite), then a large value of the interference comes from just one transmitter with an enormously high power; consequently, the asymptotics rely only on the assumption in (7.6.1).

This gives a quite good heuristic answer to the second and the third above questions. Going through the details of the proof of Theorem 7.6.1 would give some more precise ideas about some details of the questions. However, the proof in [GanTor08] fails to give a sharp answer for the third question in the case $\gamma > 1$, as it does not make assertions for the conditional measure $\mathbb{P}(\cdot \mid I > s)$. A sharp answer could be of the form that, under this conditional measure, the empirical measure of the involved devices (i.e., of those whose powers are on the scale $\log s$) converges towards some particular probability measure, and additional limit laws for their powers. Such a result could be possibly formulated and proved in terms of an LDP in the spirit of Lemma 7.3.1 for a version that also involves the powers of these devices, as in Exercise 7.3.4. This is left open in [GanTor08].

Exercise 7.6.2

Formulate a conjecture as to where in the vicinity of the origin the single high-power transmitter is located for $\gamma < 1$ and give a heuristic for the proof of this conjecture. *Hint:* For $\ell(x) = \mathbb{1}\{x \in D\}$ for some compact set $D \ni o$, there is no advantageous position within D. ◊

7.7   LDP in the Ergodic Limit

We now turn to an LDP for an object that turned out in ▶ Sect. 6.6 to play a decisive rôle in the ergodic theorem in the thermodynamic limiting setting: the empirical stationary field defined in (6.7.2). Indeed, we will formulate and discuss an LDP for $\mathcal{R}_W^{(s)}(\Phi) = |W|^{-1} \int_W dx \, \delta_{\theta_x(\Phi^{(w)})}$, for $\Phi = \sum_{i \in I} \delta_{(X_i, M_i)}$ a stationary and independently marked PPP in $\mathbb{R}^d$, in the limit $W \uparrow \mathbb{R}^d$. In ▶ Sect. 7.8 we will demonstrate how to employ such an LDP for analyzing frustration events in a telecommunication application. The following is taken from [DeuStr89, GeoZes93, DemZei10].

We recall the situation of ▶ Sect. 6.6. Let $\Phi = \sum_{i \in I} \delta_{(X_i, M_i)}$ be a stationary and i.i.d. marked PPP in $\mathbb{R}^d$ in the sense of Definition 2.4.1. We denote by $\mathcal{M}$ its mark space, which we want to assume to be a Polish space equipped with the corresponding Borel σ-algebra $\mathcal{B}(\mathcal{M})$ and a positive finite measure m. Hence, the intensity measure of Φ is equal to $\lambda \, \text{Leb} \otimes \text{m}$, for some $\lambda \geq 0$, and we can assume m to be a probability measure.

We apply the shift operator θ_x with respect to $x \in \mathbb{R}^d$ to marked points, to marked point clouds, to sets of marked point clouds and so on. We denote the set of shift-invariant probability measures on $\mathbb{S}(\mathbb{R}^d \times \mathcal{M})$ by $\mathcal{P}_\theta(\mathbb{S}(\mathbb{R}^d \times \mathcal{M}))$, i.e., the set of (distributions of) stationary marked point processes. We equip it with the tame topology, which is the one that is induced by integrals against all tame local functions, see Definition 6.1.10. Then $\Phi \in \mathcal{P}_\theta(\mathbb{S}(\mathbb{R}^d \times \mathcal{M}))$, and Φ is also ergodic. According to Corollary 6.7.3, $\mathcal{R}^{(s)}_{W_n}(\Phi)$ converges towards Φ tamely for the boxes $W_n = [-n, n]^d$. In this section, we explain the corresponding LDP for this limiting setting. This will turn out to be a far-reaching extension of Sanov's theorem, Theorem 7.2.9. Since taking Palm measures is a continuous operation, we will also obtain an LDP for the empirical individual field defined in (6.7.4) via the contraction principle, Theorem 7.2.18.

We have to introduce the function that will turn out to be the rate function. This will be defined on the set $\mathcal{P}_\theta(\mathbb{S}(\mathbb{R}^d \times \mathcal{M}))$. Having Remark 6.3.6 in mind, $\mathcal{R}^{(s)}_W(\Phi)$ is a very general version of the empirical measure of many i.i.d. random variables. Further, having Sanov's theorem in mind, we expect that the rate function for $(\mathcal{R}^{(s)}_{W_n}(\Phi))_{n \in \mathbb{N}}$ should be something like an entropy with respect to the reference measure, the distribution of Φ. This is indeed true in some sense, but there is an additional complication coming from the process level of the limiting statement: it is not possible to express the rate function as an entropy right away, rather one has to define it via a rescaled limit of entropies in large boxes relative to the volume of the box.

The restriction of a probability measure $P \in \mathcal{P}_\theta(\mathbb{S}(\mathbb{R}^d \times \mathcal{M}))$, in the first component, to a box $W \subset \mathbb{R}^d$ is denoted by P_W. More precisely, P_W is the image of P under the projection $\Pi_W \colon \phi \mapsto \phi((W \times \mathcal{M}) \cap \cdot)$. Written in terms of sets, this map is $\Pi_W(\{(x_i, m_i) \colon i \in I\}) = \{(x_i, m_i), \colon i \in I, x_i \in W\}$. For a box $W \subset \mathbb{R}^d$ we now write $H_W(P|Q)$ for the relative entropy of a probability measure $P \in \mathcal{P}(\mathbb{S}(W \times \mathcal{M}))$ on $\mathbb{S}(W \times \mathcal{M})$, with respect to another one, Q. Indeed, analogously to Eq. (7.3.1), we write

$$H_W(P|Q) = \begin{cases} \int f \log f \, dQ, & \text{if } f = dP/dQ \text{ exists,} \\ +\infty & \text{otherwise,} \end{cases} \qquad P, Q \in \mathcal{P}(\mathbb{S}(W \times \mathcal{M})).$$

When it comes to (approximate) evaluations, it may be helpful to recall the representation of H_W as a Legendre transform according to Lemma 7.2.15.

We will denote the distribution of our reference process, the homogeneous i.i.d. marked PPP Φ, by $\mathbb{Q}$ on $\mathbb{S}(\mathbb{R}^d \times \mathcal{M})$. Then we define the *specific relative entropy* with respect to $\mathbb{Q}$ by

$$h(P|\mathbb{Q}) = \begin{cases} \lim_{n \to \infty} |W_n|^{-1} H_{W_n}(P_{W_n}|\mathbb{Q}_{W_n}), & \text{if } P \in \mathcal{P}_\theta(\mathbb{S}(\mathbb{R}^d \times \mathcal{M})), \\ +\infty & \text{otherwise,} \end{cases} \qquad (7.7.1)$$

where $W_n = [-n, n]^d$, and we recall that $\mathcal{P}_\theta$ is the set of shift-invariant probability measures. Some important properties of this function are in order:

Lemma 7.7.1 (Properties of the Specific Relative Entropy) *The limit in* (7.7.1) *exists, and the function* $P \mapsto h(P|\mathbb{Q})$ *is lower semicontinuous in the tame topology, its level sets* $\{P \colon h(P|\mathbb{Q}) \leq \alpha\}$ *are compact, and* $P \mapsto h(P|\mathbb{Q})$ *is even affine, i.e., linear on the line between any two measures in* $\mathcal{P}_\theta(\mathbb{S}(\mathbb{R}^d \times \mathcal{M}))$.

Exercise 7.7.2
Show that $P \mapsto h(P|\mathbb{Q})$ is affine. *Hint:* Show convexity and concavity.   ◊

Let us for a minute neglect the space and consider only the marks by projecting on the mark space using the projection $\Pi_\mathcal{M} \colon \mathbb{R}^d \times \mathcal{M} \to \mathcal{M}$, $\Pi_\mathcal{M}((x, m_x)) = m_x$. We use the same symbol to denote the corresponding projection on point processes. Then $\Pi_\mathcal{M}(\Phi) = \sum_{i \in I} \delta_{M_i}$ is a point process in $\mathcal{M}$, and the projection of the empirical field, $|W_n|^{-1} \sum_{i \in I} \delta_{M_i} \mathbb{1}\{X_i \in W_n\}$ (called the *empirical mark field*), is nothing but an empirical measure of a large number of i.i.d. random marks each with distribution m. Their number is a Poisson variable with parameter $\lambda|W_n|$, hence a variant of Lemma 7.3.1 tells us that the empirical mark field satisfies an LDP on $\mathcal{M}$ with scale $|W_n|$ and rate function equal to $H_\mathcal{M}(\cdot|m)$, the relative entropy with respect to m. This simple observation can be turned into a formula for the specific relative entropy that might be helpful when trying to evaluate such entropies.

Remark 7.7.3 (Entropy Under Change of Mark Measure) We write $\mathbb{Q}(m)$ for the distribution of a stationary i.i.d. marked PPP with intensity measure m. Then, for any probability measure n on the mark space $\mathcal{M}$, and for any $P \in \mathcal{P}_\theta(\mathbb{S}(\mathbb{R}^d \times \mathcal{M}))$ with $P \circ \Pi_\mathcal{M}^{-1} = $ n,

$$h\big(P|\mathbb{Q}(\lambda \text{Leb} \otimes \text{m})\big) = h\big(P|\mathbb{Q}(\lambda \text{Leb} \otimes \text{n})\big) + H_\mathcal{M}(\text{n}|\text{m}). \tag{7.7.2}$$

Furthermore,

$$H_\mathcal{M}(\text{n}|\text{m}) = \inf_{P \in \mathcal{P}_\theta(\mathbb{S}(\mathbb{R}^d \times \mathcal{M})) \colon P \circ \Pi_\mathcal{M}^{-1} = \text{n}} h\big(P|\mathbb{Q}(\lambda \text{Leb} \otimes \text{m})\big), \tag{7.7.3}$$

in words, the entropy of n relative to m is equal to the smallest specific relative entropy with respect to our reference process, taken over all shift-invariant marked point processes with empirical mark distribution equal to n. This infimum is attained for $P = \mathbb{Q}(\lambda \text{Leb} \otimes \text{n})$, our reference process. The proof of these two statements is in [GeoZes93, Section 5.2].   ◊

We can now formulate the main result of this section.

Theorem 7.7.4 (LDP for the Stationary Empirical Field, [GeoZes93])
Let $\Phi = ((X_i, M_i))_{i \in I}$ be a stationary PPP in $\mathbb{R}^d$ with i.i.d. marks and $\mathbb{Q}$ its distribution. If $W_n = [-n, n]^d$, then the empirical stationary field $(\mathcal{R}_{W_n}^{(s)}(\Phi))_{n \in (0, \infty)}$ satisfies, as $n \to \infty$, an LDP in the tame topology on $\mathcal{P}(\mathbb{S}(\mathbb{R}^d \times \mathcal{M}))$ with speed $|W_n|$ and rate function $h(\cdot|\mathbb{Q})$ given in (7.7.1).

A variant of this theorem is proved in Section 5.4 of [DeuStr89] in $d = 1$ for processes Φ that are, on one hand, much more general in the sense that they are assumed to be stationary and to satisfy a strong mixing condition, but on the other hand, more restrictive since they are assumed to be continuous paths instead of being point processes.

Exercise 7.7.5 (LDP for the Empirical Field)
As one can prove with the help of Exercise 6.7.2 and the concept of exponential equivalence (see Exercise 7.2.29), the same LDP as in Theorem 7.7.4 is satisfied by the empirical field $\mathcal{R}_{W_n}(\Phi)$ defined in (6.4.7). ◇

Comments on the Proof of Theorem 7.7.4
We do not give a full proof of Theorem 7.7.4, as it is technical and lengthy, but we will give a guidance to the general strategy, following Section 5.4 in [DeuStr89] for a closely related result. Actually, we describe how to derive the LDP for the empirical field rather than for its stationary variant (which is no loss of generality because of Exercise 7.7.5).

Step 1: *Gärtner–Ellis approach for projections for the proof of the upper bound in (7.2.2).*
Fix any centered box $W \subset \mathbb{R}^d$ and prove that, for any bounded and continuous test function $F \colon \mathbb{S}(W \times \mathcal{M}) \to \mathbb{R}$, the limit

$$
\Lambda_W(F) = \lim_{n \to \infty} |W_n|^{-1} \log \mathbb{E}\Big[\exp\Big\{ \int_{W_n} F\big(\Pi_W(\theta_x(\Phi))\big)\, dx \Big\} \Big]
$$
$$
= \lim_{n \to \infty} |W_n|^{-1} \log \mathbb{E}\Big[\exp\Big\{ |W_n| \langle F, \mathcal{R}_{W_n}(\Phi) \circ \Pi_W^{-1} \rangle \Big\} \Big]
$$

(7.7.4)

exists, where we recall that Π_W is the projection operator on the box W. This mainly relies on subadditivity, but in Step 3 below, one has to go a bit deeper into this in order to have some control on the limit. According to the upper-bound half of the Gärtner–Ellis theorem (Theorem 7.2.21), we have now that the upper bound in (7.2.2) holds with rate function equal to the Legendre–Fenchel transform Λ_W^* of Λ_W.

Step 2: *Projective limit.* Now use the Dawson–Gärtner theorem (Theorem 7.2.25) to obtain that the upper bound in (7.2.2) also holds for the unprojected empirical measure, i.e., for $\mathcal{R}_{W_n}(\Phi)$ instead of $\mathcal{R}_{W_n}(\Phi) \circ \Pi_W^{-1}$, with rate function given by

$$
I(P) = \sup\{\Lambda_W^*(P \circ \Pi_W^{-1}) \colon W \subset \mathbb{R}^d \text{ a centered box}\}, \quad P \in \mathcal{P}_\theta(\mathbb{S}(\mathbb{R}^d \times \mathcal{M})).
$$

(7.7.5)

One technical step here is to show that Λ_W enjoys a certain tightness property and therefore I is a good rate function, i.e., that all its level sets $\{P \colon I(P) \leq A\}$ for $A \in \mathbb{R}$ are compact.

Step 3: *Identification of the rate function as an entropy.* Here one proves that $I(\cdot)$ is identical to the relative specific entropy $h(\cdot|\mathbb{Q})$ given in (7.7.1). In order to see that $I(P) \geq h(P|\mathbb{Q})$, one first gives an upper bound of the form

$$\Lambda_W(F) \leq \gamma_W^{-1} \log \mathbb{E}\Big[\exp\Big\{ \gamma_W \langle F, \mathcal{R}_W(\Phi) \rangle \Big\} \Big],$$

with some suitable $\gamma_W \sim |W|$ as $W \uparrow \mathbb{R}^d$. Hence, the limit of Step 1 is upper bounded against an exponential expectation with respect to a bounded function. Now using Lemma 7.2.15 (representation of $h(\cdot|\mathbb{Q})$ as a Legendre transform), we obtain '$\geq$' since I is the projective limiting Legendre transform of Λ_W. With this, also using Step 1 and Step 2, we conclude the proof of the upper bound in (7.2.2) with rate function $h(\cdot|\mathbb{Q})$. Let us note that also the converse inequality is correct, i.e., $I(P) \leq h(P|\mathbb{Q})$, which will also come as a byproduct of the following Step 4. Nevertheless, we indicate an independent proof. We use the shift invariance of Φ to upper bound, for any $P \in \mathcal{P}_\theta(\mathbb{S}(\mathbb{R}^d \times \mathcal{M}))$ and any $n > 0$, with some suitable bounded continuous function F_n,

$$\int F(\Pi_W(\phi)) \, P(\mathrm{d}\phi) - |W_n|^{-1} \log \mathbb{E}\Big[\exp\Big\{ \int_{W_n} F\big(\Pi_W(\theta_x(\Phi))\big) \, \mathrm{d}x \Big\} \Big]$$

$$\leq |W_n|^{-1}\Big[\int F_n\big(\Pi_{W+W_n}(\phi)\big) \, P(\mathrm{d}\phi) - \log \mathbb{E}\Big[\exp\big\{ F_n\big(\Pi_{W+W_n}(\Phi)\big)\big\} \Big] \Big]$$

$$\leq |W_n|^{-1} H_{W+W_n}(P_{W+W_n}|\mathbb{Q}_{W+W_n}),$$

using again the representation of the entropy as a Legendre transform. Now some elementary arguments suffice to derive that $I(P) \leq h(P|\mathbb{Q})$.

Step 4: *Proof of the lower bound in (7.2.1).* Here we fix an open set G in $\mathcal{P}_\theta(\mathbb{S}(\mathbb{R}^d \times \mathcal{M}))$ and pick any $P \in G$ and show that

$$\liminf_{n \to \infty} |W_n|^{-1} \log \mathbb{Q}\big(\mathcal{R}_{W_n}(\Phi) \in G\big) \geq -h(P|\mathbb{Q}). \tag{7.7.6}$$

This proof essentially follows an abstract version of the method that uses an exponential tilting, i.e., the Cramér transform, which we briefly outlined in Remark 7.1.6, together with the representation of the entropy as a Legendre transform, see Lemma 7.2.15. In other words, it adapts the proof of the Gärtner–Ellis theorem, Theorem 7.2.21. However, there is a technical point that we would like to point out: in the proof, it is used that the adapted exponential transformation converges towards the measure P that we picked in the beginning, but arguments for this are only available if P is not only stationary, but even ergodic. This makes a preparatory step necessary, an approximation of P with some ergodic member of $\mathcal{P}_\theta(\mathbb{S}(\mathbb{R}^d \times \mathcal{M}))$. There are some procedures known for doing that; one of them uses the ergodic decomposition theorem and the fact that $h(\cdot|\mathbb{Q})$ is affine.

So far, we have outlined the proof of [DeuStr89, Section 5.4] for a version of the LDP of Theorem 7.7.4 for hypermixing processes on the line (i.e., in $d = 1$) in the topology that is governed by continuous and bounded functions, in order to give the reader some guidance. However, the actual proof of the theorem in [GeoZes93] uses at some places other arguments, and it exclusively relies on deriving the logarithmic asymptotics of exponential expectations. Furthermore, it has to adapt the technicalities to the tame topology. We decided to omit these details.   □

We would like to stress that LDPs of the type in Theorem 7.7.4 (i.e., for the empirical measures of a stochastic process) are valid only under rather strong mixing conditions. We put the strongest assumptions in this respect by taking a PPP with independent marks.

As an easy consequence of the LDP in Theorem 7.7.4 and the contraction principle in Theorem 7.2.18, we obtain an LDP also for the empirical individual field $\mathcal{R}_{W_n}^{(s),o}(\Phi)$ defined in (6.7.4). The reason is that it is just the Palm measure of $\mathcal{R}_{W_n}^{(s)}(\Phi)$ (see Remark 6.7.6) and that taking Palm measures is a continuous operation, when we equip $\mathcal{P}(\mathbb{S}(\mathbb{R}^d \times \mathcal{M}))$ with the tame topology and $\mathcal{P}^o(\mathbb{S}(\mathbb{R}^d \times \mathcal{M}))$ with the local topology, see Lemma 6.7.5. This directly implies the following.

Corollary 7.7.6 (LDP for the Empirical Individual Field, [GeoZes93])   *Let* $\Phi =$ $((X_i, M_i))_{i \in I}$ *be a stationary PPP in* $\mathbb{R}^d$ *with independent marks. If* $W_n = [-n, n]^d$ *then the empirical individual field* $(\mathcal{R}_{W_n}^{(s),o}(\Phi))_{n \in (0,\infty)}$ *satisfies, as* $n \to \infty$, *an LDP in the local topology on* $\mathcal{P}(\mathcal{M} \times \mathbb{S}(\mathbb{R}^d \times \mathcal{M}))$ *with speed* $|W_n|$ *and rate function* h^o *given by*

$$h_{\mathbb{Q}}^o(R) = \begin{cases} h(P|\mathbb{Q}) & \text{if } R = P^o \text{ for a (necessarily unique) } P \in \mathcal{P}_\theta(\mathbb{S}(\mathbb{R}^d \times \mathcal{M})), \\ +\infty & \text{otherwise.} \end{cases}$$

(7.7.7)

The rate function $h_{\mathbb{Q}}^o$ *has compact level sets in the local topology, and it is affine.*

As another corollary, one can derive an LDP for a version of $\mathcal{R}_{W_n}(\Phi)$ that is analogous to the way in which $\mathcal{R}_{W_n}^{(s),o}(\Phi)$ is a version of $\mathcal{R}_{W_n}^{(s)}(\Phi)$ (but we cannot speak of a Palm version here).

Exercise 7.7.7 (LDP for the Canonical Empirical Individual Field)
Similar to Exercise 7.7.5, one can prove with the help of Exercise 6.7.2 and the concept of exponential equivalence (see Exercise 7.2.29) that the same LDP as in Corollary 7.7.6 is satisfied by the *canonical empirical individual field,* $\mathcal{R}_{W_n}^o(\Phi)$, where

$$\mathcal{R}_W^o(\phi) = \frac{1}{|W|} \sum_{i \in I \,:\, x_i \in W} \delta_{(m_i, \theta_{x_i}(\phi))}, \qquad \phi = ((x_i, m_i))_{i \in I}.$$

(7.7.8)

Using the contraction principle for the projection on the second component, an LDP for the empirical measure of the $\theta_{x_i}(\Phi)$, $i \in I$, follows easily.   ◊

7.8   Connectivity to Receivers in Large Boxes

Here we consider an interesting large deviation result in the model that we discussed in a broad view in ▶ Sect. 6.8 with respect to the ergodic theorem. The object of our interest here is the average local configuration of successful receivers around the transmitters (under consideration of interference), a information that contains other characteristic quantities, like the average number of possible receivers within a given distance to the transmitter, and the like. Here we are working in the ergodic limiting setting (or thermodynamic limiting setting) that we discussed in ▶ Sect. 6.4 with respect to limits and in ▶ Sect. 7.7 in view of large deviations. We will be able to apply the LDP of Corollary 7.7.6, but there are some conceptual and some technical problems to be overcome. The following is taken from [HJKP16].

Imagine $\Phi = (X_i)_{i \in I}$ and $\Psi = (Y_j)_{j \in J}$ are two independent standard PPPs in $\mathbb{R}^d$ with possibly different intensities λ and μ respectively. Each X_i has a message that would like to be transmitted to any of the Y_j. That is, Φ is the point process of the transmitters, and Ψ one of the receivers. We recall from Example 6.6.7 that the joint process (Φ, Ψ) may be seen as a standard PPP Ξ with intensity $\lambda + \mu$ with i.i.d. marks 't' and 'r' that mark each point as transmitter or receiver with probabilities $\lambda/(\lambda + \mu)$ and $\mu/(\lambda + \mu)$, respectively. In other words,

$$\Phi = \{x : (x, \mathsf{t}) \in \Xi\} \quad \text{and} \quad \Psi = \{y : (y, \mathsf{r}) \in \Xi\},$$

and in this sense we want to conceive (Φ, Ψ). We assume that all the message transmissions (better: transmission attempts) are carried out at the same time instant, and we consider here only one such instant. The success of each transmission is measured in terms of the signal-to-interference ratio as introduced in ▶ Sect. 5.2. The interference is caused by the sending attempts of all the $X_i \in \Phi$. One can easily add features into the model like random transmission powers, independently over the transmitters, or an additional noise term, and we admit them all to be present, but we do not want to write down details here.

Pick some threshold $\tau \in (0, \infty)$ and let

$$\Psi^{(i)} = \{Y_j \in \Psi : \text{SINR}(X_i, Y_j, \Phi) \geq \tau\}$$

denote the subset of receivers in Ψ to which $X_i \in \Phi$ successfully transmits. Then, we are interested in the statistical behavior of the sets $\Psi^{(i)} - X_i$ with $i \in I$ and X_i in the large observation window $W_n = [-n, n]^d$; see the illustrations and comments that we made in Example 6.8. For this, we are interested in the empirical measure,

$$L_n = L_n(\Xi) = \frac{1}{|W_n|} \sum_{i \in I : X_i \in W_n} \delta_{\Psi^{(i)} - X_i} \in \mathcal{M}(\mathbb{S}(\mathbb{R}^d)),$$

which is a random measure on the space of (receiver) configurations $\mathbb{S}(\mathbb{R}^d)$. Note that L_n is asymptotically normalized to the intensity parameter λ of the PPP Φ, i.e., $\lim_{n\to\infty} L_n(\mathbb{S}(\mathbb{R}^d)) = \lambda$, almost surely. In words, $L_n(\phi) = L_n(\Xi)(\phi)$ is, for any configuration ϕ, the average number of occurrences of ϕ centered around some transmitter in W_n as the set of reachable receivers, under the assumption that all the transmitters transmit.

We want to consider the limit as $n \to \infty$, the *thermodynamic limit*. In ▶ Sect. 6.8, we discussed the large-n limit in the light of the ergodic limit for $L_n(A)$ for a given set A of configurations, and here we consider the corresponding LDP for the random measure L_n. A comparison to the definition of an empirical canonical individual field suggests that one should aim for an application of the LDP of Exercise 7.7.7. However, it is not straightforward to conceive L_n as a functional of the canonical empirical individual field of the i.i.d. marked PPP Ξ, since (1) the correct marks $\Psi^{(i)}$ of the points X_i are still missing, (2) they are far from being i.i.d., and (3) L_n is an empirical measure only of the points in Φ, not of the points in Ψ.

In order to solve this problem, we now present a way to write L_n as a functional of the canonical empirical individual field $\mathcal{R}^o_{W_n}(\Xi)$ defined in (7.7.8).

1. For any i, add the point cloud $\Psi^{(i)}$ as an additional mark to the transmitters X_i in Ξ. This is done by the map

$$\mathcal{A}\colon \mathbb{S}(\mathbb{R}^d \times \{\mathrm{t}, \mathrm{r}\}) \to \mathbb{S}(\mathbb{R}^d \times \mathbb{S}(\mathbb{R}^d)), \qquad \mathcal{A}(\xi) = ((x_i, \psi^{(i)}))_{i\in I},$$

where we write $(x_i)_{i\in I}$ for the transmitters in ξ, i.e., $\{x\colon (x, \mathrm{t}) \in \xi\}$, and $\psi = \{y\colon (y, \mathrm{r}) \in \xi\}$ for the receivers in ξ.

2. Project the marked point configuration in $\mathbb{S}_o(\mathbb{R}^d \times \mathbb{S}(\mathbb{R}^d))$ (the set of marked configurations that have a point at the origin o) to its mark component at the origin o, i.e., use the canonical projection

$$\Pi\colon \mathbb{S}_o(\mathbb{R}^d \times \mathbb{S}(\mathbb{R}^d)) \to \mathbb{S}(\mathbb{R}^d), \qquad \Pi\big(((x_i, \psi^{(i)}))_{i\in I}\big) = \psi^{(o)},$$

where we use $o \in I$ as the index of the particle at o.

The first step transforms point clouds with marks indicating transmitters and receivers into point clouds of transmitters with marks indicating their reachable receivers. The second step singles out the configuration of reachable receivers of a transmitter at the origin. Now L_n can be expressed as follows.

Exercise 7.8.1

Work out the details and convince yourself that $L_n(\Xi) = T(\mathcal{R}^o_{W_n}(\Xi))$, where $T(P) = P(\{\mathrm{t}\} \times \cdot) \circ \mathcal{A}^{-1} \circ \Pi^{-1}$.

Hint: Use that $\delta_x \circ f^{-1} = \delta_{f(x)}$ for any map f and for any x in its domain. ◇

Now, in order to obtain an LDP for L_n, the idea is to apply the contraction principle to the map T with respect to the field $\mathcal{R}^o_{W_n}(\Xi)$ as defined in Exercise 7.7.7. Hence, the only thing that has to be done is to argue that the map T is continuous. This map consists of first evaluating the process on transmitters, then taking the image measure with respect to $\mathcal{A}$, followed by taking the image measure with respect to Π. The first and the last maps are easily seen to be continuous.

But the continuity of the second one is more tricky; more precisely, it is strictly speaking not true, but can be approximately derived in a technical way. Carrying out this is precisely the main point in the proof of the following theorem, which is the main result of [HJKP16, Theorem 1]. Recall the definition of the relative entropy density h given in (7.7.1). We denote by $\mathbb{Q}$ the distribution of the above reference process Ξ, i.e., a point process on $\mathbb{R}^d \times \{t, r\}$, conceived as a process of independent transmitters and receivers (Φ, Ψ).

Theorem 7.8.2   (LDP for Connectable Receivers, [HJKP16])

The (distributions of the) random measures L_n satisfy the LDP with rate $|W_n|$ and rate function given by

$$\mathcal{M}_1\big(\mathcal{M}(\mathbb{S}(\mathbb{R}^d))\big) \ni Q \mapsto \inf_{R \in \mathcal{P}:\, R(\{t\}\times\cdot)\circ\mathcal{A}^{-1}\circ\Pi^{-1}=Q} h^o_{\mathbb{Q}}(R)$$

$$= \inf_{P \in \mathcal{P}_\theta:\, P^o(\{t\}\times\cdot)\circ\mathcal{A}^{-1}\circ\Pi^{-1}=Q} h(P|\mathbb{Q}).$$

Comments on the Proof

As mentioned above, the main point is an application of the contraction principle for the function T, as defined in Exercise 7.8.1, to the LDP in Exercise 7.7.7 for the empirical canonical individual field. As we explained above, the point is the continuity of the map that takes the image measure with respect to the map $\mathcal{A}$. This continuity is to be understood in the local topology. There are three obstacles here: the missing unboundedness of the support of the path-loss function, the lack of continuity of the interference in the denominator of the definition of the SINR, and the lack of continuity of the event that some random variable (here the SINR) is larger than the given threshold τ. Each of these obstacles is only of technical nature and can be overcome by approximations: cutting the path-loss function to finite interaction length, bounding the number of points Y_j in the vicinity of a given X_i, and replacing the strict inequality '$> \tau$' by the inequality '$\geq \tau - \varepsilon$' for some $\varepsilon > 0$. Each of these three steps is carried out using the concept of exponentially good approximations, which we discussed at the end of ▶ Sect. 7.2.   □

Maybe it is helpful to envision the rate function: for a given Q, a distribution of a positive measure on receiver point configurations in a neighborhood of the origin, one can lower bound the value of the rate function in Q by the entropy $h(P|\mathbb{Q})$ of any shift-invariant point process with respect to (Φ, Ψ), which has the property that the application of the map T of Exercise 7.8.1 to this point process leads to Q; the best lower bound that is obtained in this way is the value of the rate function. An evaluation of the rate function is (as usual) a very hard analytical problem since we are dealing with huge spaces of stationary (marked) point processes.

Random Malware Propagation: Interacting Markov Processes

In contrast to the preceding chapters, where we considered the propagation of *wanted* messages over the telecommunication system, we now turn to the propagation of *unwanted* ones, e.g., of malware. We introduce, on the set $\Phi = (X_i)_{i \in I}$ of all the device locations, a *time-dependent* Markovian model of the set of infected devices, evolving in time as a stochastic process. The main random mechanism that drives this infection process has two elements: (1) each device that has an infected neighbor is also infected after an exponentially distributed random time, and (2) any infected device either remains infected, or it undergoes a spontaneous healing, or it becomes patched by some neighboring device of another type, again after independent exponentially distributed random times.

This is a well-known mechanism of a Markovian random process on a state space of the form E^Φ. One famous example is the *contact process* where every device X_i assumes at any time one of the two possible states in $E = \{$infected, susceptible$\}$, and it may randomly change its state over time according to the above rules. The contact process is a prominent representative in the theory of *interacting particle systems* (IPS). In a nutshell, this is the theory of continuous-time Markov jump processes on discrete configuration spaces of the form E^Φ, where Φ carries some neighboring structure, see [Lig85, Lig13, Dur88] for standard textbooks on the subject. Traditionally IPSs are studied on fixed geometries like lattices $\mathbb{Z}^d$ or trees, but in recent years, first results became available, for example, for the contact process on a homogeneous PPP Φ in $\mathbb{R}^d$, see [MenSin16].

The spontaneous healing can be seen as a *counter measure* that the operator has installed in the system against the propagation of the malware. Although well studied in the literature, we consider such a model not as the most suitable one for describing conceivable counter measures in ad-hoc telecommunication networks. Rather, we introduce a third possible state in the model, which can, for example, describe a state of immunity against the malware or even a state that carries some vaccine or *goodware* designed to permanently remove the malware. This gives rise to new or at least much less investigated propagation models.

In ▶ Sect. 8.1 we introduce the (well-known) theory of continuous-time Markov chains on finite sets and explain how jump rates characterize the process. In ▶

© Springer Nature Switzerland AG 2020
B. Jahnel, W. König, *Probabilistic Methods in Telecommunications*, Compact Textbooks in Mathematics,
https://doi.org/10.1007/978-3-030-36090-0_8

Sect. 8.2 we start to study the first relevant model of malware spreading without counter measures, the *Richardson model*. Here we also introduce the fundamental notions of invariant measures and ergodicity for an IPS. We present a shape theorem on $\mathbb{Z}^d$ as well as existence and phase transitions for the infection speed of Richardson models on Poisson–Gilbert graphs, see Definition 4.3.1. In ▶ Sect. 8.3 we add to the Richardson model a first simple healing mechanism, which gives rise to the well-known *contact process*. Here we recall the famous phase transition of survival of the infection on $\mathbb{Z}^d$ and also present the associated shape theorem before we partially reproduce these result for the contact process on Poisson–Gilbert graphs. Finally, in ▶ Sect. 8.4 we consider models in which another, more appropriate, counter measure is used. More precisely, another device type is introduced, the white knights, which can vaccinate infected devices but not susceptible ones. This gives rise to a picture of chase and escape for the goodware against the malware. Again we start by stating known results about the model on fixed geometries before presenting first results on the Poisson–Gilbert graph.

8.1   Markov Chains in Continuous Time

Let us explain how a Markov chain in continuous time on a discrete state space is characterized and constructed generally in terms of waiting times and jump rates. This is just a short summary of a fundamental theory, which is discussed at length in many textbooks, for example in [And12, Lig85, Lig10]. We do this here only for a finite set Ω as the state space.

Let $(\omega(t))_{t\in[0,\infty)}$ denote a stochastic process (i.e., a family of random variables, indexed by $t \in [0, \infty)$) on the finite set Ω. It is called a *Markov chain* if

$$\mathbb{P}(\omega(t_{n+1}) = a_{n+1} \mid \omega(t_1) = a_1, \ldots, \omega(t_n) = a_n)$$

$$= \mathbb{P}(\omega(t_{n+1}) = a_{n+1} \mid \omega(t_n) = a_n),$$

for all $n \in \mathbb{N}, a_1, \ldots, a_{n+1} \in \Omega$ and $0 \leq t_1 < \cdots < t_{n+1}$. We call it *time-homogeneous* if the transition probabilities

$$\mathbb{P}(\omega(t) = a \mid \omega(s) = b) = p_{t-s}(a, b), \qquad a, b \in \Omega, 0 \leq s < t, \tag{8.1.1}$$

depend on the time differences only (and of course on the states).

We assume that $(\omega(t))_{t\in[0,\infty)}$ is a separable[1] time-homogeneous Markov chain such that the transition probabilities in (8.1.1) are continuous in t at $t = 0$[2]. Now we will describe its probabilistic structure. For any state $a \in \Omega$, we write $\mathbb{P}^a$ for the probability measure if the process starts in a under $\mathbb{P}^a$, i.e., $\mathbb{P}^a(\omega(0) = a) = 1$, and we call a its initial state.

---

[1] The definition of separability of a Markov chain is very technical; it ensures that the paths $t \mapsto \omega(t)$ are not too 'wild'. We decided not to give this definition here.

[2] Then the collection of these transition probabilities is called *standard*.

Since Ω is finite and the index set an interval, $(\omega(t))_{t \in [0,\infty)}$ must be necessarily a *jump process*. In order to avoid pathologies, we assume that, with probability one under any $\mathbb{P}^a$, there are infinitely many finite jump times $0 < \tau_1 < \tau_2 < \cdots < \infty$ such that $\lim_{n \to \infty} \tau_n = \infty$, i.e., we exclude infinitely many jumps in finite time (so-called explosions). Let $\tau_1 = \inf\{t > 0 : \omega(t) \neq \omega(0)\}$ denote the first jump time, i.e., the first time at which the particle leaves its initial state. Then, by the Markov property and time homogeneity,

$$\mathbb{P}^a(\tau_1 > t + h | \tau_1 > t) = \mathbb{P}^a(\tau_1 > t + h | \omega_t = a) = \mathbb{P}^a(\tau_1 > h), \qquad t, h \in (0, \infty).$$

Hence, $\mathbb{P}^a(\tau_1 > t+h) = \mathbb{P}^a(\tau_1 > h)\mathbb{P}^a(\tau_1 > t)$, but this is a functional equation that can only be solved by the exponential function. Thus, $\mathbb{P}^a(\tau_1 > t) = \exp(-\lambda_a t)$, for some $\lambda_a \in (0, \infty)$, which means that τ_1 must be exponentially distributed. The parameter λ_a can be recovered via

$$-\frac{d}{dh}\bigg|_{h=0} \mathbb{P}^a(\tau_1 > h) = \lambda_a.$$

The decision of the process at the jump times is described by a stochastic matrix $p = (p_{a,b})_{a,b \in \Omega}$, where

$$p_{a,b} = \mathbb{P}^a(\omega(\tau_1) = b) \in [0, 1]$$

is the probability to jump from state a to state b at a time of a jump. Since we consider only proper jumps, we have $p_{a,a} = 0$. Then p is the transition matrix of a discrete-time Markov chain that never stands still. We introduce the *transfer matrix* (or *Q-matrix* or *generator*) of the process,

$$L = (c(a, b))_{a,b \in \Omega} \qquad \text{where } c(a, b) = -\frac{d}{dh}\bigg|_{h=0} \mathbb{P}^a(\omega_h = b). \tag{8.1.2}$$

Then outside the diagonal, the Q-matrix contains the *jump rates* of the process, while its row sums are equal to zero:

$$c(a, b) = \begin{cases} \lambda_a p_{a,b} \in [0, \infty) & \text{if } a \neq b, \\ -\lambda_a = -\sum_{d \in \Omega \setminus \{a\}} c(a, d) & \text{if } a = b. \end{cases} \tag{8.1.3}$$

The fact that the row sums of L are equal to zero is easy to guess, since L is the time derivative at time zero of the transition matrix, whose row sums are equal to one. Since this property of L is not self-evident if Ω is infinitely large, one calls L *conservative* if it is non-negative off the diagonal and has all the row sums equal to zero.

Certainly, the further propagation of the Markov process X is done by an extension of the above described procedure, which also takes into account the second jump and so on. In this way, one in particular proves the *strong Markov property* at these jump times in an explicit way, and it turns out that all the differences between subsequent jump times as well as all the jump decisions are independent. Alternatively, one proves the strong Markov property for general stopping times, which requires some technical work.

In this way, we arrive now at a good understanding of the probabilistic mechanism of a time-homogeneous, separable Markov chain with standard transition probabilities.

Remark 8.1.1 (Construction of the Markov Chain) A complete construction of the chain $(\omega(t))_{t \in [0,\infty)}$ is given by the following recipe. Given a state $a \in \Omega$ and a time $t \in [0, \infty)$, then, on the event $\{\omega(t) = a\}$, the further evolution is described by generating an exponential random variable τ with parameter λ_a and an independent Ω-valued random variable J with distribution $p(a, \cdot)$, independently of the entire past $(\omega(s))_{s \in [0,t)}$. Then the process is defined on $[t, t + \tau]$ by putting

$$
\omega(s) = \begin{cases} \omega(t) & \text{if } s \in (t, t + \tau), \\ J & \text{if } s = t + \tau. \end{cases}
$$

Iterating this procedure infinitely often produces the entire Markov chain.

Remark 8.1.2 (Alternative Construction) Instead of constructing first the jump time variables and then the jump decision variables as in Remark 8.1.1, one can alternatively construct the Markov chain as follows. Again, given a state $a \in \Omega$ and a time $t \in [0, \infty)$, then, on the event $\{\omega(t) = a\}$, for any $b \in \Omega \setminus \{a\}$, we choose independent exponentially distributed random variables $\sigma_{a,b}$ with parameter $c(a, b)$, independently of all the past. Then the index b such that the variable $\sigma_{a,b}$ elapses first is taken as the site J where the chain jumps to at time $t + \sigma_{a,b}$, that is, $\tau = \min_{b \in \Omega \setminus \{a\}} \sigma_{a,b}$ is the jump time.

Exercise 8.1.3 (Equivalence of the Constructions)
Prove that the two constructions given in Remarks 8.1.1 and 8.1.2 are equivalent to each other and that either of them produces a Markov chain that satisfies (8.1.2).

Jointly with the above constructions comes the important theoretical fact that the transfer matrix L, together with the initial state (or an initial distribution), characterizes the distribution of the process completely. More precisely, for any separable and time-homogeneous Markov chain on Ω with standard transition probabilities, L exists as in (8.1.2), and the jump time differences and all the jump decisions are independent random variables with the distribution that is described in Remarks 8.1.1 or 8.1.2. These constructions show also the converse: given a starting probability vector and a conservative matrix L, there is a time-homogeneous separable Markov chain $(\omega(t))_{t \in [0,\infty)}$ on Ω such that (8.1.2) holds.

A common way to describe the jump mechanism is in terms of the asymptotic

$$
\mathbb{P}^\nu(\omega_{t+h} = b \mid \omega_t = a) = c(a, b)h + o(h), \qquad h \downarrow 0, \tag{8.1.4}
$$

for $a \neq b$ and initial distribution ν on Ω and transfer matrix L.

Exercise 8.1.4

Verify that the description (8.1.4) is correct.

The paths of the Markov chain $(\omega(t))_{t \in [0,\infty)}$ are right-continuous everywhere and constant strictly between the jump times and have left limits at the jump times. In other words, the paths lie in the space of *càdlàg paths*,

$$D[0, \infty) = \{f : [0, \infty) \to \Omega : f \text{ is right-continuous and has left limits everywhere}\}.$$

To fix some measurable structure on $D[0, \infty)$, we at least want to be able to evaluate the process at any given time on the Boreal σ-algebra $\mathcal{B}(\Omega)$ on Ω, i.e., we equip $D[0, \infty)$ with the σ-algebra generated by the events $\{\omega_t \in A\}$ for all $t \geq 0$ and $A \in \mathcal{B}(\Omega)$. A natural filtration $(\mathcal{F}_t)_{t \in [0,\infty)}$ is then given by $\mathcal{F}_t = \sigma(\omega_s : s \in [0, t])$.

Example 8.1.5 (Poisson Process)

In the case where $\Omega = \mathbb{N}_0$, $\omega(0) = 0$ and $c(a, b) = q(\mathbb{1}\{b = a + 1\} - \mathbb{1}\{b = a\})$ for all $a, b \in \mathbb{N}_0$, the process of jump times is a PPP on $[0, \infty)$ with intensity q. The Markov process $(\omega(t))_{t \in [0,\infty)}$ increases by one after independent exponential times with parameter q. Such a process can be used as the counting process for another (time-homogeneous) Markov chain in continuous time; it 'rings' at all times at which that process makes a jump. ◊

So far, we considered continuous-time Markov chains only on a finite set. However, in our telecommunication applications in ▶ Sects. 8.2–8.4, we will actually consider Markov chains on uncountable state spaces, namely on state spaces of the form $\Omega = E^\phi$, where E is a finite set and $\phi = (x_i)_{i \in I} \in \mathbb{S}(\mathbb{R}^d)$ is an infinite but locally finite point configuration in $\mathbb{R}^d$. The standard examples are $\phi = \mathbb{Z}^d$ or the d-ary tree $\phi = \mathbb{T}^d$, but we will also take ϕ as the Poisson–Gilbert graph that we know from Definition 4.3.1. We assign to each x_i a property $\omega(t, x_i) \in E$ at time t, then $\omega(t) = (\omega(t, x))_{x \in \phi} \in E^\phi$ is the state of the Markov chain at time t. In our models, we will be given explicit rates on the configuration space, and we want to make assertions about the corresponding Markov chain. Two serious mathematical problems in its construction arise from the infiniteness of ϕ. First, a priori there can be infinitely many positive rates for the change of the state of some $x \in \phi$; therefore, one has to rule out the possibility that the jump decision is made immediately. Second, a priori a jump decision may depend on infinitely many other sites.

These problems are already present in models with non-random ϕ. Here conditions can be formulated that ensure well-definedness of the Markov chain. This is closely related to the theory of Markov generators and their associated semigroups, more precisely the Hille–Yosida theory. In the standard reference [Lig85], for processes on *fixed* geometries ϕ, e.g., $\phi = \mathbb{Z}^d$, such conditions are given in terms of uniform summability and locality properties of the jump rates, but we do not want to enter into details here; rather refer to [Lig85, Chapter 1]. A Markov chain satisfying these condition is called an *interacting particle system (IPS)* in the sense of [Lig85]. We will adopt this terminology for all the models that we are going to discuss. Note that in the simplest non-trivial case, the jump rates are only for the change of the state of a single site, and the rates only depend on a bounded neighborhood of the location of the change. For sufficiently regular fixed ϕ, this important subclass of Markov chains is well-defined.

However, already in our standard example of a random geometry, where $\phi = \Phi$ is a PPP with the neighborhood structure of the Poisson–Gilbert graph for some distance parameter $R > 0$, the conditions as presented in [Lig85] are usually violated almost surely in Φ. To see this, note that in the Poisson–Gilbert graph the sequence of the degrees of the nodes is unbounded with probability one. Hence, if, for example, the jump rate of a site is linear in the number of its neighbors, there is no bound on the jump rates uniformly in the sites. Fortunately, for the models we consider in the sequel, we do not have to develop extensions of the IPS theory based on generators as in [Lig85], but rather we can resort to an alternative and more direct construction, which is closely related to the ideas of Remarks 8.1.1 and 8.1.2, the *graphical representation*. Here, by our choice of the initial configuration, the first jump will always only concern the single site at the origin, which resolves the issue that there should not be an immediate jump at time zero. The well-definedness for all positive times will be an easy consequence of the results that we present in the following sections. For the time being, we will not enter these details, but take the existence of the Markov chain for granted.

Let us introduce some notation that we will use in all the examples of IPS we consider in this chapter. The transition rates $c(\zeta, \xi)$ will be defined for pairs of configurations $\zeta, \xi \in \Omega = E^\phi$ with $\phi = (x_i)_{i \in I}$ a locally finite point configuration in $\mathbb{R}^d$ and E being either $\{0, 1\}$ or $\{0, 1, 2\}$. We will keep things simple (and more natural) and require that the only steps that the Markov chain can make are flips of the property at some site x_i from one to another state. Hence, the only non-zero rates $c(\zeta, \xi)$ will be those in which ξ differs from ζ in precisely one entry, i.e., the jumps from ζ to the configuration $\zeta^{(i,e)} = (\zeta^{(i,e)}(x))_{x \in \phi}$ defined by

$$\zeta^{(i,e)}(x) = \begin{cases} \zeta(x) & \text{if } x \in \phi \setminus \{x_i\}, \\ e & \text{if } x = x_i. \end{cases}$$

Hence, only rates $c(\zeta, \zeta^{(i,e)})$ with $\zeta(x_i) \neq e$ have to be considered for $i \in I$ and $e \in E$. That is, the Markov chain proceeds at time t by changing the value of $\omega(t, x_i)$ to e after independent exponential times with parameter $c(\omega(t), \omega^{(i,e)}(t))$. These rates $c(\zeta, \zeta^{(i,e)})$ will be local in the sense that they depend only on $\zeta(y)$ with sites y that are neighboring to x_i. In the case of $\phi = \Phi$ a PPP, we will rely on the Poisson–Gilbert graph structure on ϕ, where two sites in ϕ are neighbors if their distance is smaller than a parameter R. Depending on the situation that is to be modeled, one could also replace the Poisson–Gilbert graph with the Poisson–SINR graph that we discussed in ▶ Sect. 5.3 and also change the underlying point process to be, for example, a Cox point process instead of a PPP, see ▶ Chap. 3, giving rise to Cox–Gilbert graphs. Except for the results mentioned in the introduction of this chapter, there are no specific results in the literature for IPSs on these types of graphs yet.

The three models that we consider in ▶ Sects. 8.2–8.4 are, in a nutshell, characterized as follows. In the Richardson models of ▶ Sect. 8.2, the malware grows unboundedly by contact with infected devices. In the contact processes in ▶ Sect. 8.3, there is an additional, simple counter measure that spontaneously reboots devices. Finally, in the chase-escape models in ▶ Sect. 8.4, the countermeasure consists of patching infected devices via neighboring devices which carry the software patch. In particular, in all the models under consideration, any change of the state (i.e., any jump in the Markov chain) in our telecommunication system corresponds either to a message transmission (an infection with a malware or a transmission of a patch from neighbor to neighbor) or to a reboot of the device. All these jumps happen after exponential waiting times that are independent of everything that happened before, but their parameters depend on the current state of the system.

Remark 8.1.6 (Non-exponential Waiting Times) Here is a fundamental criticism about the use of exponentially distributed random variables for modeling waiting times before a change in the telecommunication system. In the context of malware propagation in telecommunication systems, this modeling assumption is hard to justify for various reasons, starting with the fact that transmissions need to have, for technical reasons, a certain deterministic delay. The use of exponential waiting times is mainly driven by the desire to have the Markov property at disposal, which makes many aspects much easier, and to be able to work with an explicit formula for the distribution of the waiting times. However, for example for simulations, it is often no extra effort to use other types of waiting time distributions, for example a uniform distribution on the interval $[c, d]$ with $0 < c < d$. Developing a theory for IPSs based on non-exponential waiting times is a challenging but important aspect in the mathematical analysis of telecommunication networks.

8.2 Richardson Models

In this section, we present a first non-trivial step towards an application of continuous-time Markov chains to the modeling of the random propagation of malware over a system of devices. The best well-known process that describes this is the *Richardson model*, an IPS that we introduce now. In this process each particle can have only the two properties 'susceptible' ('0') or 'infected' ('1'), and there are only flips from 0 to 1 (representing an infection), and the infection rate is linear in the number of infected neighbors.

The Richardson model [Ric73] is characterized by the following jump rates on $\Omega = \{0, 1\}^\phi$, where $\phi = (x_i)_{i \in I}$ is a locally finite point configuration with graph structure in $\mathbb{R}^d$:

$$c(\zeta, \zeta^{(i)}) = \lambda_{\mathrm{I}} \sum_{j \in I : x_j \sim x_i} \mathbb{1}\{\zeta(x_j) = 1\}, \text{ if } \zeta(x_i) = 0,$$

where $\sim$ denotes the neighborhood relation on ϕ, and we write $\zeta^{(i)} = \zeta^{(i,1)}$ for the unique configuration that sets $\zeta^{(i,1)}(x_i) = 1$ and otherwise does not change ζ. All other rates $c(\zeta, \xi)$ are equal to zero, i.e., only single-site states are changed in a jump and only previously healthy sites can become infected. In this process, a healthy particle becomes infected with the rate equal to λ_I times the number of infected neighbors. It is thus a pure growth model. The parameter $\lambda_I > 0$ tunes the strength of the infection. If there is no infected particle in the system, there will never be any infected particle entering. The model has a natural interpretation as to represent a random spread of malware in a system of nodes ϕ equipped with a graph structure allowing us to speak of neighbors. There is no possibility of recovery or immunization in this model. The Richardson model is a prototypical example of an *attractive spin system*, which means in simple words that configurations with more infected devices lead to faster infections of healthy devices, for details see [Lig85, Chapter 1].

The analysis of an IPS typically starts with the consideration of *invariant measures*, i.e., distributions of initial configurations that remain invariant under the stochastic time evolution, just like fixed points of an ordinary differential equation. To define this properly, let $\mathbb{P}^\nu$ denote the measure under which the IPS is defined with the initial distribution ν on Ω. We use the measure-theoretic notation $\mathbb{P} \circ X^{-1}$ for the distribution of a random variable X under a probability measure $\mathbb{P}$.

Definition 8.2.1 (Invariant Measures and Ergodic IPS)

A probability measure ν on Ω is called *invariant* for the IPS if

$$\mathbb{P}^\nu \circ \omega(t)^{-1} = \nu, \qquad \text{for any } t \geq 0.$$

Moreover, the IPS is called *ergodic* if it has a unique invariant measure ν that satisfies, for all starting distributions ν',

$$\lim_{t \to \infty} \mathbb{P}^{\nu'} \circ \omega(t)^{-1} = \nu \qquad \text{weakly.}$$

The set of invariant measures is always non-empty and a convex set, and hence it can be represented by the subset of extremal invariant measures, which are the ones that cannot be represented by a convex combination of other invariant measures, see [Lig85, Chapter 1].

In the case of the Richardson model on the deterministic grid $\phi = \mathbb{Z}^d$, equipped with the usual neighborhood relation, it is clear that δ_0 and δ_1, the measures which put mass one on the all-zero respectively the all-one configuration, are invariant for any choice of $\lambda_I > 0$. In particular, the process is not ergodic. It is a general fact of attractive systems that the limiting measure of the process when started in δ_0 and δ_1 are extremal invariant measures [Lig85, Theorem 2.3. Chapter 2]. In particular for the Richardson model, δ_0 and δ_1 are the only extremal invariant measures. Moreover, it is evident that for any starting configuration that is different from the all-zero configuration, the Richardson model on $\mathbb{Z}^d$ converges to δ_1, since it is a pure growth model.

Exercise 8.2.2

What is the structure of the invariant measures for the Richardson model on ϕ, a realization of a PPP on $\mathbb{R}^d$ with edge structure given via the Poisson–Gilbert graph, see Definition 4.3.1?

The analysis of invariant measures for the Richardson model is straightforward and features no critical behavior, in contrast to the contact process presented in ▶ Sect. 8.3. This is of course due to the simplicity of the updating. Nevertheless, further interesting questions can be studied, for instance about the limiting shape of the set of infected devices when the malware is initially present only at a fixed set of devices, for example, only the origin. Let us present here the corresponding result for $\phi = \mathbb{Z}^d$. For this let

$$I(t) = \{x \in \phi : \omega(t, x) = 1\}$$

denote the set of infected devices at time t. In order to state the theorem, we also need a continuous-space version of $I(t)$, more precisely we define $I_c(t) = \bigcup_{x \in I(t)} (x + (-1/2, 1/2]^d)$. Let δ_o denote the Dirac measure which deterministically puts an infection only in the origin.

Theorem 8.2.3 (Shape Theorem for the Richardson Model on $\mathbb{Z}^d$)
Let $\phi = \mathbb{Z}^d$, then there exists a non-random compact convex set $B \subset \mathbb{R}^d$, which is invariant under permutation of and reflection in the coordinate hyperplanes and has a nonempty interior, such that for any $\lambda_I > 0$ and any $\varepsilon > 0$, the event

$$\left\{ \text{there exists } T < \infty \text{ such that for all } t > T : (1 - \varepsilon)B \subseteq \frac{I_c(t)}{\lambda_I t} \subseteq (1 + \varepsilon)B \right\}$$

has probability one under the measure $\mathbb{P}^{\delta_o}$.

This theorem can be found, for example, in [CoxDur81, HägPem00]. To determine the exact shape of the set B is still an open problem, mainly due to the non-isotropy of $\mathbb{Z}^d$. From the theorem we can in particular see that the growth of the infected set is asymptotically linear in λ_I and in time. As in the case of the shape theorem for the Boolean model, Theorem 4.6.7, the proof of Theorem 8.2.3 is based on a version of Kinsman's subadditivity ergodic theorem.

In the context of device-to-device networks, it is interesting to consider the Richardson model based on a *random* set of nodes Φ equipped with a neighborhood relation. A natural example is the Poisson–Gilbert graph on a standard PPP Φ. The set of infected devices $I(t)$ then has two sources of randomness, namely the random graph and the IPS evolution on the random graph. For random graphs that are isotropic in distribution, the distribution of $I(t)$ is also isotropic if $I(0)$ is isotropic, see ◻ Fig. 8.1 for an illustration based on a CPP.

In particular any limiting set B as in Theorem 8.2.3 should be a ball with some radius α, where α can be interpreted as the speed of the infection spread. To the best of our knowledge, a result corresponding to Theorem 8.2.3 in this random setting is not yet

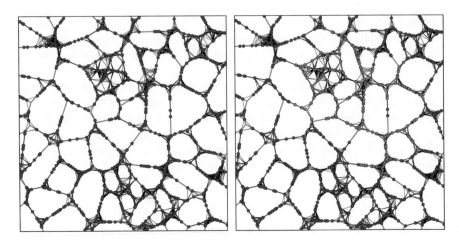

◻ Fig. 8.1 Realization of randomly placed devices on a street system of PVT type. Left: initial time with one infected user in red at the centre. Right: some finite-time snapshot in which the malware has started to infect surrounding devices

available in the literature. However, let us present here a first step in that direction. We consider the Richardson model under $\mathbb{P}^{o,\delta_o}$, which is given as a two-layer process

$$\mathbb{E}^{o,\delta_o}[f(\omega)] = \mathbb{E}^o\big[\mathbb{E}^{\delta_o}[f(\omega)|\Phi]\big],$$

for any bounded measurable function f on trajectories of configurations, where $\mathbb{P}^o$ denotes the Palm measure of a homogeneous PPP Φ with intensity $\lambda > 0$ and $\mathbb{P}^{\delta_o}(\cdot|\Phi)$ denotes the Richardson model on Φ equipped with the edges given by the Poisson–Gilbert graph with parameter $R > 0$ and with initial infection only at the origin, which is present in Φ with probability one under $\mathbb{P}^o$. Further let

$$\tau_u = \inf\{t > 0 \colon I(t) \not\subset B_u(o)\}$$

denote the hitting time of the infection of the boundary of a ball around the origin with given radius u, and define the asymptotic speed of the process as

$$\limsup_{u\uparrow\infty} \mathbb{E}^{o,\delta_o}[u/\tau_u] = \alpha \in [0, \infty]. \qquad (8.2.1)$$

Recall that there are three parameters in the model: the density λ of the PPP, the radius parameter R for the Poisson–Gilbert graph, and the infection speed parameter λ_I. Hence, α should be seen as a function of these three.

Exercise 8.2.4
Show the following scale invariance for the asymptotic speed of the Richardson model on the Poisson–Gilbert graph:

$$b\alpha(b^d \lambda, b^{-1} R, \lambda_I) = \alpha(\lambda, R, \lambda_I) = b^{-1}\alpha(\lambda, R, b\lambda_I),$$

for any $b > 0$.

Recall from ▶ Sect. 4.3 the super- and subcritical percolation regimes of the Gilbert graph. We have the following theorem.

Theorem 8.2.5 (Infection Speed for the Richardson Models on Poisson–Gilbert Graphs)
Consider the Richardson model on a Poisson–Gilbert graph based on a stationary PPP with parameters λ_I, λ and R. Then, we have $\alpha < \infty$. Moreover, $0 = \alpha$ if and only if the Poisson–Gilbert graph is in the subcritical percolation regime.

The proof actually shows that, in the supercritical regime,

$$\frac{\theta(\lambda, R)\lambda_I}{\rho(\lambda, R)} \leq \alpha \leq \frac{R\lambda_I}{c(\gamma)}, \tag{8.2.2}$$

where $\theta = \theta(\lambda, R)$ is the percolation probability, see (4.3.2), ρ is the stretch factor, see Theorem 4.6.7, and $c(\gamma)$ is the solution to the fixed point equation $\log(\gamma x) = x - 1$ in $(0, 1/\gamma)$, where $\gamma = \lambda|B_R(o)|$ is the connective constant in the Boolean model according to Theorem 4.6.11.

Sketch of Proof
For simplicity, write $\mathbb{P}^{o,\delta_o} = \mathbb{P}$ and $\mathbb{E}$ for the corresponding expectation. First we consider the subcritical regime. Note that for $R_t = \inf\{u > 0 : I(t) \subset B_u(o)\}$, the smallest radius of a centered ball containing the infected sites, we can equivalently write $\mathbb{P}(R_t \geq u) = \mathbb{P}(I(t) \not\subset B_u(o)) = \mathbb{P}(\tau_u \leq t)$. By $\mathcal{C}_R(o)$ we denote the cluster containing the origin of the Poisson–Gilbert graph. Note that for all $t \geq 0$, $R_t \leq \#\mathcal{C}_R(o)R$, where we write $\#\mathcal{C}_R(o)$ for the number of nodes in $\mathcal{C}_R(o)$. Hence,

$$\limsup_{u\uparrow\infty} \mathbb{E}[u/\tau_u] = \limsup_{t\uparrow\infty} \mathbb{E}[R_t/t] = \limsup_{t\uparrow\infty} \frac{1}{t}\sum_{m\in\mathbb{N}} \mathbb{E}[\mathbb{1}\{\#\mathcal{C}_R(o) = m\}R_t]$$

$$\leq \limsup_{t\uparrow\infty} \frac{R}{t}\sum_{m\in\mathbb{N}} m\mathbb{P}(\#\mathcal{C}_R(o) = m) = \limsup_{t\uparrow\infty} \frac{R}{t}\mathbb{E}[\#\mathcal{C}_R(o)].$$

Now use that in the subcritical regime $\mathbb{E}[\#\mathcal{C}_R(o)] < \infty$, see part 5 of Remark 4.5.1 to deduce that $\alpha = 0$.

For the supercritical regime, let us start by writing

$$\mathbb{E}[u/\tau_u] = \int \mathrm{d}t\, \mathbb{P}(u/\tau_u \geq t) = \int \mathrm{d}t\, \mathbb{P}(\tau_u \leq u/t). \tag{8.2.3}$$

We first derive an upper bound. For this, denote by $K^n(\Phi)$ the set of self-avoiding paths of length n starting from the origin in the Poisson–Gilbert graph Φ (see the text prior to Theorem 4.6.11). Further, denote by $\tau_u(\kappa)$ the first time that the malware (i.e., the process

$(I(t))_{t\in[0,\infty)})$ hits $B_u(o)^c$ along the path $\kappa \in K^n(\Phi)$. Then we have for any $c > 0$ that

$$\mathbb{P}(\tau_u \le cu) = \mathbb{E}^o\big[\mathbb{P}^{\delta_o}(\tau_u \le cu \,|\, \Phi)\big] = \mathbb{E}^o\Big[\mathbb{P}^{\delta_o}\Big(\bigcup_{n\in\mathbb{N}_0}\bigcup_{\kappa\in K^n(\Phi)}\{\tau_u(\kappa) \le cu\}\,\Big|\,\Phi\Big)\Big]$$

$$\le \sum_{n\ge u/R}\mathbb{E}^o\Big[\sum_{\kappa\in K^n(\Phi)}\mathbb{P}^{\delta_o}(\tau_u(\kappa) \le cu \,|\, \Phi)\Big],$$

where, in the third step, we use that in order to reach $B_u(o)^c$ at least u/R steps are required. On the event $\{\tau_u(\kappa) \le cu\}$, the sum of n independent exponential waiting times $T_1, \ldots, T_n$ with parameter λ_I is smaller than cu. We use the Markov inequality for the exponential function with parameter $a > 0$ (similarly to (7.1.1)) and write

$$\mathbb{P}^{\delta_o}(\tau_u(\kappa) \le cu \,|\, \Phi) \le \mathbb{P}^{\delta_o}\Big(\sum_{i=1}^n T_i \le cu\Big) \le e^{acu}\Big(\frac{\lambda_I}{\lambda_I + a}\Big)^n.$$

Now, recalling the connective constant $\gamma = \lambda|B_R(o)|$ for the Boolean model from Theorem 4.6.11 (which, in this particular case, gives the exact expected cardinality of $K^n(\Phi)$), we have, as long as $\gamma\lambda_I/(\lambda_I + a) < 1$, the estimate

$$\mathbb{P}(\tau_u \le cu) \le e^{acu}\sum_{n\ge u/R}\Big(\gamma\frac{\lambda_I}{\lambda_I + a}\Big)^n \le (1 - \gamma\frac{\lambda_I}{\lambda_I + a})^{-1}e^{acu}\Big(\gamma\frac{\lambda_I}{\lambda_I + a}\Big)^{u/R}$$

$$= \text{const.} \times \exp\Big\{u\Big[ac + \frac{1}{R}\log\frac{\gamma\lambda_I}{\lambda_I + a}\Big]\Big\}.$$

Note that in the supercritical regime we have in particular $\gamma > 1$, and the inequality $\gamma\lambda_I/(\lambda_I + a) < 1$ is only satisfied if $a > \lambda_I(\gamma - 1)$. Hence, we can minimize the sum in the exponent over all the a's and achieve that in the exponent we have a negative sign. Taking the derivative with respect to a, we see that the global minimum is achieved in $a = (cR)^{-1} - \lambda_I$ if and only if $c < (R\lambda_I\gamma)^{-1}$. For these choices of a and c the inequality

$$1 - R\lambda_I c + \log(\gamma R\lambda_I c) < 0$$

is not satisfied in general. However, the fixed point equation $\log(\gamma x) = x - 1$ has a solution in $(0, 1/\gamma)$, which we denote $c(\gamma)$, and the inequality is satisfied for $c < c(\gamma)(R\lambda_I)^{-1}$.

Using this estimate, we can now come back to (8.2.3) and bound the speed from above by

$$\mathbb{E}[u/\tau_u] \le \frac{R\lambda_I}{c(\gamma)} + \frac{1}{1 - \gamma c(\gamma)}\int_{R\lambda_I/c(\gamma)}^{\infty} dt \exp\Big(\frac{u}{R}\Big(1 - \frac{R\lambda_I}{t} + \log\frac{\gamma R\lambda_I}{t}\Big)\Big)$$

$$\le \frac{R\lambda_I}{c(\gamma)} + \frac{1}{1 - \gamma c(\gamma)}\exp\Big(\frac{u}{r}\big(1 - c(\gamma) + \log(\gamma r\lambda_I)\big)\Big)\int_{R\lambda_I/c(\gamma)}^{\infty} dt\, t^{-\frac{u}{R}}.$$

For $u > R$ we can then calculate

$$\int_{R\lambda_I/c(\gamma)}^{\infty} dt\, t^{-\frac{u}{R}} = \frac{R}{u-R}\left(\frac{R\lambda_I}{c(\gamma)}\right)^{1-\frac{u}{r}},$$

and hence, by the definition of $c(\gamma)$,

$$\mathbb{E}[u/\tau_u] \leq \frac{R\lambda_I}{c(\gamma)} + \frac{R^2\lambda_I}{c(\gamma)(u-R)(1-\gamma c(\gamma))} \exp\left(\frac{u}{R}\big(1-c(\gamma)+\log(\gamma c(\gamma))\big)\right)$$

$$= \frac{R\lambda_I}{c(\gamma)} + \frac{R^2\lambda_I}{c(\gamma)(u-R)(1-\gamma c(\gamma))}.$$

Since the second summand tends to zero as u tends to infinity, we arrived at the upper bound for α in (8.2.2) in particular $\alpha < \infty$.

For proving the lower bound in (8.2.2), we proceed similarly. We have

$$\mathbb{P}(\tau_u \leq cu) \geq \mathbb{E}^o\left[\mathbb{1}\{o \leftrightsquigarrow \infty\}\mathbb{P}^{\delta_o}\left(\bigcup_{n\geq u/R}\bigcup_{\kappa\in K^n(\Phi)} \{\tau_u(\kappa)\leq cu\}\,\Big|\,\Phi\right)\right],$$

using that in order to reach $B_u(o)^c$ at least u/R steps are required, and we add the indicator on the event $\{o \leftrightsquigarrow \infty\}$ that the origin is connected to ∞.

Now, under the event $\{o \leftrightsquigarrow \infty\}$, there exists a path κ connecting o to infinity. Moreover, using the stretch factor $\rho > 0$, see Theorem 4.6.7, we can pick the self-avoiding path κ' in such a way that the number of hops in κ' required to reach $B_u(o)^c$ is bounded from above by $(\rho+\varepsilon)u$, for all sufficiently large u. In particular, for all sufficiently large u, we have

$$\mathbb{E}^o\left[\mathbb{1}\{o \leftrightsquigarrow \infty\}\mathbb{P}^{\delta_o}\left(\bigcup_{n\geq u/R}\bigcup_{\kappa\in K^n(\Phi)} \{\tau_u(\kappa)\leq cu\}\,\Big|\,\Phi\right)\right]$$

$$\geq \mathbb{E}^o\left[\mathbb{1}\{o \leftrightsquigarrow \infty\}\mathbb{P}^{\delta_o}(\tau_u(\kappa')\leq cu|\Phi)\right] \geq \theta\mathbb{P}^{\delta_o}\left(\sum_{i=1}^{(\rho+\varepsilon)u} T_i \leq cu\right)$$

$$= \theta\left(1 - \mathbb{P}^{\delta_o}\left(\sum_{i=1}^{(\rho+\varepsilon)u} T_i > cu\right)\right) \geq \theta\big(1 - \exp(-acu)\mathbb{E}[\exp(aT_1)]^{(\rho+\varepsilon)u}\big),$$

where in the last line we again use the Markov inequality for the exponential function with parameter $a > 0$. Note again that $\mathbb{E}[\exp(aT_1)] = \lambda_I/(\lambda_I - a)$ and hence we must pick $a > 0$ and $c > 0$ in such a way that

$$ac - (\rho+\varepsilon)\log\frac{\lambda_I}{\lambda_I - a} > 0.$$

Taking the derivative with respect to a, we see that the optimal choice for a is given by $a = \lambda_I - (\rho+\varepsilon)/c$ given that $c > (\rho+\varepsilon)/\lambda_I$, which leads to the requirement that

$$(\rho+\varepsilon)\log\frac{c\lambda_I}{\rho+\varepsilon} - c\lambda_I + (\rho+\varepsilon) < 0.$$

The left-hand side is an entropy and thus the inequality is always satisfied for $c > (\rho+\varepsilon)/\lambda_I$. We have just performed some large deviations estimate explicitly.

To finish the proof, observe that for $c > \rho/\lambda_I$ we now have for the speed the lower bound

$$\mathbb{E}[u/\tau_u] \geq \mathbb{E}[\mathbb{1}\{\tau_u \leq cu\}u/\tau_u] \geq c^{-1}\mathbb{P}(\tau_u \leq cu) \geq c^{-1}\theta\big(1 - c' \exp(-c''u)\big).$$

for some constants $c', c'' > 0$. Letting u tend to infinity and optimizing over c, we have

$$\liminf_{u\uparrow\infty}\mathbb{E}[u/\tau_u] \geq \theta\lambda_I/\rho,$$

where θ is the percolation probability, see (4.3.2). This finishes the proof.   □

Exercise 8.2.6 (Uniform Waiting Times)

Recall from Remark 8.1.6 that non-exponential waiting times may be more suitable for modeling the delay in transmissions. Show, for all the waiting times being independently uniformly distributed on $[c, d]$ with $0 < c < d$, that

$$\theta(\lambda, R)\frac{1}{\rho(\lambda, R)d} \leq \liminf_{u\uparrow\infty}\mathbb{E}^{o,\delta_o}[u/\tau_u] \leq \limsup_{u\uparrow\infty}\mathbb{E}^{o,\delta_o}[u/\tau_u] \leq \theta(\lambda, R)\frac{1}{\rho(\lambda, R)c}. \quad \Diamond$$

8.3   Contact Processes

We now consider an extension of the Richardson model for modeling the random propagation of malware in a device-to-device network. Here, infected devices can become spontaneously healthy but not immune, i.e., susceptible again. This is the famous *contact process*, see for example [Lig13]. In the context of malware propagation, we interpret the spontaneous healing as a rebooting of devices and thereby elimination of the malware. This is our first mechanism of a countermeasure against the malware; in ▶ Chap. 8.4 we will consider a more sophisticated one.

The contact process is again defined on $\Omega = E^\phi$ for $E = \{0, 1\}$, '1' standing for 'infected' and '0' for 'susceptible', and a general countably-infinite site space ϕ with some neighborhood structure. As before, the most important examples are $\phi = \mathbb{Z}^d$ and $\phi = \Phi$ a PPP with the graph structure of an associated Poisson–Gilbert graph.

The contact process is characterized by jump rates on Ω of the form $c(\zeta, \zeta^{(i)})$ with $i \in I$, where $\zeta^{(i)}$ is the unique configuration that differs from ζ precisely in x_i, with

$$c(\zeta, \zeta^{(i)}) = \begin{cases} 1, & \text{if } \zeta(x_i) = 1, \\ \lambda_I \sum_{j\in I: \, x_j\sim x_i} \mathbb{1}\{\zeta(x_j) = 1\}, & \text{if } \zeta(x_i) = 0, \end{cases} \quad (8.3.1)$$

and $\sim$ denotes the neighborhood relation. All other rates $c(\zeta, \xi)$ are equal to zero, i.e., again only single-site states are changed in a jump. The second line in (8.3.1) is the mechanism of the Richardson model (a susceptible site becomes infected with a rate that is λ_I times the number of infected neighbors), but in the first line in (8.3.1) we see that additionally an infected site can become spontaneously susceptible again with fixed rate one, independently of all the other sites. The parameter $\lambda_I > 0$ tunes the strength of the infection. Note that a susceptible site can again be infected later; there is no immunization. Hence, each particle can change its state several times. If the state of entire susceptibility is reached, the infection can never come back, i.e., the state δ_0 is an *absorbing state* and in particular invariant under the dynamics. Now, in contrast to the Richardson model, the question arises whether or not the infection can die out with positive probability or not, and on the event of eternal survival of the infection, how large the set of infected sites will be after late times and what shape it might have.

The contact process, although appearing innocent, shows many interesting features already on the lattice $\phi = \mathbb{Z}^d$ with the usual neighborhood structure. Most prominently, it has a phase transition with respect to the survival of the infection. Using the same definition of $I(t)$ as for the Richardson model in Theorem 8.2.3, we define

$$\tau = \inf\{t > 0 : I(t) = \emptyset\}, \tag{8.3.2}$$

the *extinction time* of the process. Then we say that the contact process *dies out* if

$$\mathbb{P}^{\delta_o}(\tau = \infty) = 0 \tag{8.3.3}$$

and *survives* otherwise. Note that by the invariance of the dynamics with respect to lattice translations, the statement whether or not the process dies out is invariant under shifts of the initial point where the infection starts, i.e., o can be replaced by x for any $x \in \mathbb{Z}^d$. We have the following theorem.

Theorem 8.3.1 (Phase Transition for the Contact Process on $\mathbb{Z}^d$)
Consider $\phi = \mathbb{Z}^d$, then there exists $\lambda_{I,cr} \in (0, \infty)$ such that the contact process dies out if and only if $\lambda_I \leq \lambda_{I,cr}$.

The theorem was first proved for $d = 1$, where the existence of a survival phase can even be guaranteed for a large class of attractive systems. This allows then to only consider the upper-invariant measure, which is the limiting measure for the system started in δ_1. Existence of a non-trivial extinction phase follows from general arguments about weakness of updates, see [Lig85, Chapter I]. The proof for the statement in all dimensions including the behavior at criticality (i.e., for $\lambda_I = \lambda_{I,cr}$) followed later [BezGri90], see also [Lig13].

Remark 8.3.2 (Properties of the Contact Process on $\mathbb{Z}^d$)
1. *Invariant measures.* In the survival regime, the set of invariant measures is given by convex combinations of δ_0 and another invariant measure ν_{λ_I}, usually referred to as the *upper invariant measure*. It can be defined as $\lim_{t \to \infty} \mathbb{P}^{\delta_1} \circ \omega(t)^{-1} = \nu_{\lambda_I}$. In this regime, in particular, the system is not ergodic.

2. *Complete convergence.* For $\lambda_I > \lambda_{I,cr}$ and any initial state $I(0) \subset \mathbb{Z}^d$, the contact process converges weakly, as $t \uparrow \infty$, to

$$\mathbb{P}^{\delta_{I(0)}}(\tau = \infty)\nu_{\lambda_I} + \mathbb{P}^{\delta_{I(0)}}(\tau < \infty)\delta_0,$$

where $\delta_{I(0)}$ denotes the initial measure which deterministically puts infected devices only in $I(0)$.

3. *Value of $\lambda_{I,cr}$.* Approximations for the critical rate for $d = 1$ suggest that $\lambda_{I,cr}^{(1)} \approx 1.6494$, see [Lig85, page 275]. Via coupling arguments, it is easy to see that $d \mapsto \lambda_{I,cr}^{(d)}$ is decreasing. Moreover, using a comparison with branching random walks, we have $\lambda_{I,cr}^{(d)} \geq 1/2d$ and it can be shown that $\lambda_{I,cr}^{(d)} \leq 2/d$, in fact $\lim_{d\uparrow\infty} d\lambda_{I,cr}^{(d)} = 1/2$, see [Lig85, Theorem 1.33 and 4.1].

4. *Strong survival.* The contact process is said to *strongly survive* if $\mathbb{P}^{\delta_o}\big(o \in I(t)$ for infinitely many $t\big) > 0$, which is by translation invariance a property independent of the origin o. In general, for the critical value for strong survival we have $\lambda_{I,cr}' \geq \lambda_{I,cr}$. In case of the contact process on $\mathbb{Z}^d$, a comparison of the contact process to oriented percolation can be used to show that indeed $\lambda_{I,cr}' = \lambda_{I,cr}$, see [BezGri90, Lig13]. On the other hand, for the contact process on trees with degree at least 3, we have $\lambda_{I,cr}' > \lambda_{I,cr}$, see [Pem92].

5. *Infection probability.* The map $\lambda_I \mapsto \mathbb{P}^{\delta_o}(\tau = \infty) = \nu_{\lambda_I}\big(\zeta(o) = 1\big)$, which maps the rate to the survival probability, is expected to share some similarities with the percolation probability $\lambda \mapsto \theta(\lambda)$ as a function of the intensity parameter, see ◻ Fig. 4.2. In particular, close to the corresponding critical values, both quantities should behave like a power law with some critical exponent, which is an interesting and hard open research question.

6. *Relation to Richardson model.* Obviously the Richardson model can be used to dominate the contact process, for example, with respect to the speed of infection spread, see (8.2.2).

Remark 8.3.3 (Related Models)

1. *Multi-type contact processes.* The contact process can also be generalized to model multiple competing viruses, see [Val10]. In this case, based on the initial distribution of the viruses, coexistence or extinction of one or more viruses can be observed. This is also an important research direction in the context of telecommunication systems.

2. *Voter models.* Another classical model that has an interpretation in telecommunications is the *voter model*, see [Lig85, Lig13]. Here, two different opinions, or viruses, are in the system and the updating in the simplest case is given by

$$c(\zeta, \zeta^{(i)}) = \begin{cases} \sum_{j \in I : x_j \sim x_i} \mathbb{1}\{\zeta(x_j) = 1\}, & \text{if } \zeta(x_i) = 0, \\ \sum_{j \in I : x_j \sim x_i} \mathbb{1}\{\zeta(x_j) = 0\}, & \text{if } \zeta(x_i) = 1. \end{cases}$$

In words, the configuration ζ flips site x_i with a rate equal to the number of neighboring sites that are in the opposite state. This system is symmetric in 0 and 1 and has invariant states δ_0 and δ_1 and is hence never ergodic. Apart from the analysis of existence of additional extremal invariant measures, which turn out to exist for $\phi = \mathbb{Z}^d$ with

$d \geq 3$, the analysis is concerned with coexistence and clustering of the viruses. The main technical advantage of the voter model is that it can be related via duality to coalescing random walks, see [Lig13] for details. Also extensions to multitype models with potentially infinite-range dependencies have been considered in the literature.

Also for the contact process on $\phi = \mathbb{Z}^d$, a shape theorem is available, where in contrast to the definition of $I(t)$ as used in Theorem 8.2.3, we now use

$$I'(t) = \{x \in \phi \colon \omega(s, x) = 1 \text{ for some } s \in [0, t]\},$$

the set of sites that have ever been infected by time t, and the continuum version I'_c similar to I_c. Note that for the Richardson model $I'(t) = I(t)$, but for the contact process $I'(t)$ and $I(t)$ do not necessarily coincide. In particular, in case of the contact process, $I'(t)$ is almost surely increasing in t and hence again a growth process, whereas $I(t)$ is not. The following theorem is formulated only for the supercritical regime of the one-dimensional contact process, with critical intensity defined as $\lambda_{I,cr}^{(1)} \in (0, \infty)$, see [DurGri82].

Theorem 8.3.4 (Shape Theorem for the Contact Process on $\mathbb{Z}^d$)
Let $\phi = \mathbb{Z}^d$, then there exists a non-random compact convex set $B \subset \mathbb{R}^d$, which is invariant under permutation of and reflection in the coordinate hyperplanes and has a nonempty interior, such that for any $\lambda_I > \lambda_{I,cr}^{(1)}$ and any $\varepsilon > 0$, the event

$$\left\{ \text{there exists } T < \infty \text{ such that for all } t > T \colon (1 - \varepsilon)B \subseteq \frac{I'_c(t)}{\lambda_I t} \subseteq (1 + \varepsilon)B \right\}$$

has probability one under the measure $\mathbb{P}^{\delta_o}$ conditioned on the survival of the infection.

Again the proof is based on an application of a version of the subadditivity ergodic theorem. A crucial ingredient in the proof is the fact that for $\lambda_I > \lambda_{I,cr}$ there exist constants $c, \varepsilon > 0$ such that

$$\mathbb{P}^{\delta_o}(t < \tau < \infty) \leq c \exp(-\varepsilon t), \qquad t \in [0, \infty),$$

i.e., it is exponentially unlikely to survive for a long time and still die out, see [Lig13, Theorem 2.30].

The contact process on fixed geometries is the subject of a very long list of publications, but still a number of questions are unresolved, for example concerning the behavior at the critical parameter. Much less is known about the contact process on random geometries such as Poisson–Gilbert graphs, see ◘ Fig. 8.2 for an illustration.

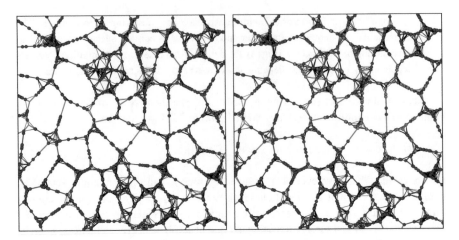

◘ Fig. 8.2 Realization of randomly placed devices on a street system of PVT type subject to the contact process. Left: initial time with one infected device in red at the centre. Right: some finite time snapshot in which the malware has started to infect surrounding devices in red but certain devices are again susceptible at the given time

As mentioned earlier, one of the challenges when dealing with the contact process on the Poisson–Gilbert graph is that nodes in the graph are of unbounded degree. More precisely, we even have

$$\mathbb{P}\big(\sup_{i \in I} \deg(X_i) = \infty\big) = 1.$$

This is not only a technical obstacle for constructing the Markov chain (see the remarks at the end of ▶ Chap. 8.1), but can potentially lead also to peculiar effects with respect to the survival of the infection, where survival is defined as in the lattice case, see (8.3.3). Indeed, in almost every configuration, occasionally the contact process might infect a node with an extremely high degree, which then is potentially able to boost the infection and hence lead to survival even for small infection rates λ_I. The following theorem, see [MenSin16, Corollary 4.2.], guarantees the existence of subcritical phases of extinction for the contact process.

Theorem 8.3.5 (Subcritical Phase for the Contact Process on Poisson–Gilbert Graphs)
Consider the contact process on the Poisson–Gilbert graph of a stationary PPP. Then there exists $0 < \lambda_{I,cr} < \infty$ such that for all $\lambda_I < \lambda_{I,cr}$ the infection dies out.

The proof is based on a layering procedure of the nodes of the underlying random graph based on their degree. This method is also of independent interest, and in particular, allows to treat many more random geometries.

Let us now provide a complementary result to Theorem 8.3.5 about survival of the infection. For this, recall the definition of the asymptotic speed of the process from (8.2.1)

$$\limsup_{u \uparrow \infty} \mathbb{E}^{o,\delta_o}[u/\tau_u] = \alpha \in [0, \infty].$$

Note that in this definition we can still use $\tau_u = \inf\{t > 0 \colon I(t) \not\subset B_u(o)\}$ since it has the same distribution as $\tau'_u = \inf\{t > 0 \colon I'(t) \not\subset B_u(o)\}$.

Theorem 8.3.6 (Positive Speed for the Contact Process on Poisson–Gilbert Graphs)

Consider the contact process on a supercritical Poisson–Gilbert graph based on a stationary PPP. Then there exists $0 < \lambda_{\mathrm{I,cr}} < \infty$ such that for all $\lambda_{\mathrm{I}} > \lambda_{\mathrm{I,cr}}$ we have that $\alpha > 0$.

Let us note that at this point it is an open research question what the relation is between survival and positivity of the asymptotic speed. Survival and extinction could imply vanishing speed as well as positivity of the speed. The following sketch of the proof of Theorem 8.3.6 is based on a comparison to the one-dimensional case.

Sketch of Proof
We proceed similar as in the proof of the lower bound in Theorem 8.2.5, but since we want to import known results about the contact process instead of performing the estimates explicitly, we have to be more careful. Again, in this proof we abbreviate $\mathbb{P} = \mathbb{P}^{o,\delta_o}$. Let us start by estimating

$$\mathbb{E}[u/\tau_u] \geq \mathbb{E}[\mathbb{1}\{o \leftrightsquigarrow \infty\}u/\tau_u] = \int \mathrm{d}t\, \mathbb{E}^o\big[\mathbb{1}\{o \leftrightsquigarrow \infty\}\mathbb{P}^{\delta_o}(\tau_u \leq u/t|\Phi)\big].$$

In the supercritical percolation regime, $\theta = \mathbb{P}^o(o \leftrightsquigarrow \infty) > 0$, and hence with positive probability, there exists a self-avoiding path κ on the Poisson–Gilbert graph that connects o to ∞. Moreover, using the stretch factor $\rho > 0$, see Theorem 4.6.7, for any $\varepsilon > 0$ and sufficiently large u, we can pick the path κ in such a way that the number of hops in κ required to reach $B_u(o)^c$ is at most $u(\rho + \varepsilon)$. Then, let $\tau_u(\kappa)$ denote the first hitting time of $B_u(o)^c$ of the contact process along the path $\kappa \subset \Phi$, neglecting all influences coming from $\Phi \setminus \kappa$, in other words, the first hitting time of the coupled process on κ. Note that the infection can not spread faster on κ than on Φ and hence,

$$\mathbb{E}^o\big[\mathbb{1}\{o \leftrightsquigarrow \infty\}\mathbb{P}^{\delta_o}(\tau_u \leq u/t|\Phi)\big] \geq \mathbb{E}^o\big[\mathbb{1}\{o \leftrightsquigarrow \infty\}\mathbb{P}^{\delta_o}(\tau_u(\kappa) \leq u/t|\Phi)\big].$$

Now we want to identify the contact process on κ with the contact process on $\mathbb{N}_0$ and import known results for that process. More precisely, we use [DurGri82, Lemma, page 546], which states that for $\lambda_{\mathrm{I}} > \lambda_{\mathrm{I,cr}}^{(1)}$, where $\lambda_{\mathrm{I,cr}}^{(1)}$ is defined as in Theorem 8.3.4, the survival probability for the contact process on $\mathbb{N}_0$ is positive, i.e., $\beta = \mathbb{P}^{\delta_o}\big(\tau(\mathbb{N}_0) = \infty\big) > 0$, and that there exist constants $c, c', \alpha' \in (0, \infty)$ such that

$$\mathbb{P}^{\delta_o}\big(\tau_u(\mathbb{N}_0) > u/t, \tau(\mathbb{N}_0) = \infty\big) \le c \exp(-c'u/t), \qquad t \le \alpha'.$$

Equipped with these statements and using the fact that

$$\tau_u(\kappa) \le \tau_{u(\rho+\varepsilon)}(\mathbb{N}_0),$$

we can further estimate,

$$\mathbb{E}[u/\tau_u] \ge (\rho + \varepsilon)^{-1} \int_0^{\alpha'} dt \, \mathbb{E}^o\big[\mathbb{1}\{o \rightsquigarrow \infty\}\mathbb{P}^{\delta_o}\big(\tau_u(\kappa) \le u(\rho + \varepsilon)/t, \tau(\kappa) = \infty|\Phi\big)\big]$$

$$\ge \frac{\theta}{\rho + \varepsilon} \int_0^{\alpha'} dt \, \mathbb{P}^{\delta_o}\big(\tau_{u(\rho+\varepsilon)}(\mathbb{N}_0) \le u(\rho + \varepsilon)/t, \tau(\mathbb{N}_0) = \infty\big)$$

$$= \frac{\theta}{\rho + \varepsilon}\alpha'\beta - \frac{\theta}{\rho + \varepsilon} \int_0^{\alpha'} dt \, \mathbb{P}^{\delta_o}\big(\tau_{u(\rho+\varepsilon)}(\mathbb{N}_0) > u(\rho + \varepsilon)/t, \tau(\mathbb{N}_0) = \infty\big)$$

$$\ge \frac{\theta}{\rho + \varepsilon}\alpha'\beta - \frac{\theta}{\rho + \varepsilon} \int_0^{\alpha'} dt \, c \exp(-c'u(\rho + \varepsilon)/t)$$

$$\ge \frac{\theta}{\rho + \varepsilon}\alpha'\beta - \frac{\theta}{\rho + \varepsilon}\alpha' c \exp(-c'u(\rho + \varepsilon)/\alpha').$$

Finally, since we can additionally require that u is sufficiently large, such that $c \exp(-c'u(\rho + \varepsilon)/\alpha') < \beta/2$ and ε is arbitrary, we have

$$\liminf_{u\uparrow\infty} \mathbb{E}[u/\tau_u] \ge \frac{\theta}{2\rho}\alpha'\beta,$$

which finishes the proof.   □

Exercise 8.3.7
Derive and prove scaling formulas similar to the ones presented in Exercise 8.2.4 for the contact process on Poisson–Gilbert graphs.   ◊

Exercise 8.3.8
Prove that for $\lambda_I < \gamma^{-1}$ we have $\alpha = 0$ where γ is the connective constant for the Boolean model, see Theorem 4.6.11. *Hint:* There is a very elegant argument using Martingale theory. Verify the stronger statement that $\mathbb{E}[I] < \infty$ in the following way. Note that the one-dimensional contact process is dominated by the voter model on $\mathbb{N}_0$, started in $I_{\text{voter}}(0) = \{o\}$ with absorbing state o. Verify that for suitable $a > 1$ and $\lambda_I < a^{-1}$, $a^{|I_{\text{voter}}(t)|}$ is a super Martingale, which implies that $\mathbb{P}^{\delta_o}(I_{\text{voter}}(\tau_n) = n + 1) \le (a - 1)/(a^{n+1} - 1)$, where $\tau_n = \inf\{t \ge 0 : I_{\text{voter}}(t) = n + 1 \text{ or } I_{\text{voter}}(t) = o\}$. Now bounding

$$\mathbb{E}[I] \le \sum_{n\in\mathbb{N}} \gamma^n \mathbb{P}^o(I_{\text{voter}}(\tau_n) = n + 1)$$

gives the result. Note that $\alpha = 0$ does not necessarily imply extinction of the malware. $\Diamond$

8.4 Chase-Escape Models

The contact process is certainly not the most accurate model for malware propagation. Modeling criticisms are manifold, starting from the Markovian assumption for the waiting times (see Remark 8.1.6), the restriction to only one virus, to the fact that a virus typically should be defeated by installing a software patch (instead of rebooting), which then makes the device immune to further attacks by this virus. In this final section, we introduce a class of IPS models that at least takes into account the last aspect, the *chase-escape models*.

The idea is that, instead of a simple rebooting, malware can only be eliminated by a patch, which has to be forcefully installed on infected devices. Due to legal regulations, this update cannot be performed on any device but only on those ones which are infected. Due to the ad-hoc nature of our system, in the initial configuration, we have to now distinguish between susceptible devices, infected devices and patch carrying devices. The last type is sometimes dubbed the *white knights*, which cannot be infected again and are hence immune. More mathematically, we consider a three-state IPS on the configuration space $\Omega = E^\phi$ with $E = \{0, 1, 2\}$. The state 0 represents susceptibility, state 1 represents infection and 2 represents immunity. Then the jump rates are defined as

$$c(\zeta, \zeta^{(i,e)}) = \begin{cases} \sum_{j \in I : x_j \sim x_i} \mathbb{1}\{\zeta(x_j) = 2\}, & \text{if } \zeta(x_i) = 1, e = 2, \\ \lambda_I \sum_{j \in I : x_j \sim x_i} \mathbb{1}\{\zeta(x_j) = 1\}, & \text{if } \zeta(x_i) = 0, e = 1, \end{cases}$$

where $\zeta^{(i,e)}$ differs from ζ only in x_i, where $\zeta^{(i,e)}(x_i) = e$. All other rates are zero. The second line is as in the Richardson model: a healthy particle is infected with rate equal to a parameter times the number of infected neighbors. The first line says that infected particles are healed and immunized with rate equal to the number of neighboring white knights. The white knights will never be changed, they forward their patch to other neighboring infected particles. Any particle can only change from susceptible to infected and from infected to immune. The parameter $\lambda_I > 0$ tunes the frequency of infection compared to immunization. This IPS is called the *chase-escape model*, since the white knights chase the infection, which tries to escape. We keep the notions and notations for the set of infected particles at time t, $I(t) = \{x \in \phi : \omega(t, x) = 1\}$, the extinction time $\tau = \inf\{t > 0 : I(t) = \emptyset\}$ and survival probability $\mathbb{P}^{\delta_o}(\tau = \infty) > 0$ assuming that $o \in \phi$.

First note that the model has the trivial invariant measures δ_0, δ_1 and δ_2 which are also absorbing states. A thorough analysis of the set of extremal invariant measures is yet an open research topic. However, first results recently became available for the question of survival of the infection for a number of particular fixed geometries, which we report here.

Let $\mathbb{T}^d$ denote the d-ary tree, the rooted tree where each vertex has precisely d children. Assume that at initial time zero there is one infected device at the origin (the root) and one white knight connected to the origin as an additional site. The following result was first proved in [Kor05].

Theorem 8.4.1 (Critical Intensities for Chase-Escape Models on $\mathbb{T}^d$)
Let $\phi = \mathbb{T}^d$ for $d \geq 2$ and $\lambda_{\mathrm{I,cr}}(d) = 2d - 1 - 2\sqrt{d^2 - d}$, then for $\lambda_{\mathrm{I}}^{(d)} \leq \lambda_{\mathrm{I,cr}}$ the infection dies out.

The above theorem provides a good example for the accuracy of the general wisdom that in the setting of trees explicit results can be obtained which are otherwise hard to prove, for example in the lattice setting. Note that, for example, $\lambda_{\mathrm{I,cr}}(2) \approx 0.17$ is substantially less than the patch rate 1. More generally, note that $\lambda_{\mathrm{I,cr}}(d) = 1/4d + o(d^{-2}) < 1$. This means that the malware can be substantially slower than the patch and still the infection survives. This is an effect coming from the boundary of $I(t)$ where roughly speaking the infection has more options to further spread than the chasing white knights. A simple proof is provided in [DJT18] and based on the reflection principle for simple random walks together with the number of self-avoiding n-step paths, which is easy to evaluate explicitly for trees. The proof in particular admits the corollary that the critical parameter for any graph with connective constant bounded by $\gamma \in \mathbb{N}$, see Theorem 4.6.11, is bounded from above by $\lambda_{\mathrm{I,cr}}(\gamma)$ for the tree $\mathbb{T}^\gamma$, see [DJT18, Corollary 2]. Further, in [DJT18], Theorem 8.4.1 is extended to the ladder graph $\phi = \mathbb{Z} \times \{0, 1\}$. For $\phi = \mathbb{Z}^d$, only partial results are available, which indicate that also in this case survival of the infection is possible for $\lambda_{\mathrm{I}} < 1$. In ◘ Fig. 8.3 we present simulation for $\phi = \mathbb{Z}^2$.

In view of our main topic, the analysis of spatial device-to-device networks, it is again desirable to study chase-escape models on random geometries. Considering the scope of this book, typical examples would be the Poisson–Gilbert graph based on homogeneous PPPs or Cox–Gilbert graphs based on CPPs, or ultimately maybe Cox-SINR models, which provide random graphs with enormously different degrees of complexity but also modeling accuracy. One possible setup would then be to again consider the Palm version of the model and start with an initial infection at the origin. In the following, we consider only the simplest case mentioned above, the Poisson–Gilbert graph based on a superposition of two homogeneous PPPs Φ_{W} and Φ_{D} representing white knights and normal devices with intensity $\lambda_{\mathrm{W}} \geq 0$ and $\lambda_{\mathrm{D}} \geq 0$, respectively. Note that equivalently, we could consider a joint PPP Φ with intensity $\lambda_{\mathrm{W}} + \lambda_{\mathrm{D}}$ and then use an i.i.d. marking D, W with Bernoulli probability with parameter $p = \lambda_{\mathrm{W}}/(\lambda_{\mathrm{W}} + \lambda_{\mathrm{D}})$, which would underline the idea that an operator might be interested in the initial proportion of devices carrying the patch. An illustration of the associated chase-escape model for CPPs is given in ◘ Fig. 8.4.

As a first rough criterion for the resilience of the device-to-device network, we can measure the ability of the infection to survive for all times in expectation over the random network. This is a so-called *annealed* approach, see Remark 8.4.5.

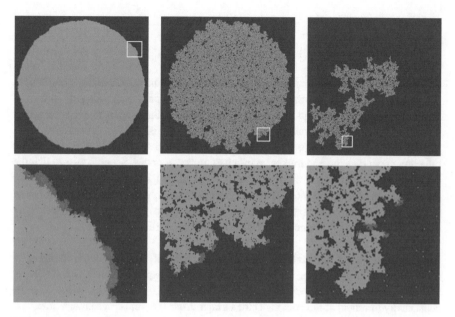

◘ Fig. 8.3 Three realizations of the chase-escape model on $\mathbb{Z}^2$ for some finite time with $\lambda_I = 1, \lambda_I = 0.55$ and $\lambda_I = 0.50$, from left to right (in the first row) with corresponding enlargements of some boundary areas (in the second row). The right-most realization shows the model close to criticality. Blue vertices are susceptible, green vertices indicate white knights, and red vertices at the boundary of the green vertices are infected

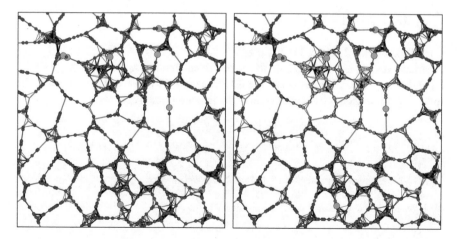

◘ Fig. 8.4 Realization of randomly placed devices on a street system of PVT type. The larger green discs indicate randomly placed white knights. Left: initial time with one infected device in red at the center. Right: some finite time snapshot in which the malware has started to infect surrounding devices in red. The green regular size devices are vaccinated by a software update initiated by some contact of the infection with a white knight

We again denote by

$$\tau = \inf\{t > 0 \colon I(t) = \emptyset\}$$

the extinction time of the infection. If the origin is disconnected from the white-knight process, then $\tau = \infty$. On the other hand, if there is a path connecting the origin to the white-knight process and the cluster of the origin is finite, then $\tau < \infty$. Hence, it is useful to distinguish two cases. First, we say that the infection *survives globally* if the event that $\tau = \infty$ and $\#(\bigcup_{t \geq 0} I(t)) = \infty$ has positive probability. In words, the infection never dies out and infects an infinite number of nodes. Second, we say that the infection *survives locally* if the event that $\tau = \infty$ and $\#(\bigcup_{t \geq 0} I(t)) < \infty$ has positive probability. In words, the infection never dies out but only infects a finite number of nodes. It is not hard to see that the infection always survives locally and in the following exercise we derive some bounds on the associated probability.

Exercise 8.4.2 (Local Survival for Chase-Escape Models on Poisson–Gilbert Graphs)
Show that in all parameter regimes of the system, $\exp(-(\lambda_{\mathrm{W}} + \lambda_{\mathrm{D}})|B_R(o)|) \leq \mathbb{P}^{o,\delta_o}$ (the infection survives locally) $\leq (1 - \theta(R, \lambda_{\mathrm{D}})) \exp(-\lambda_{\mathrm{W}}|B_R(o)|)$, where R is the connectivity threshold of the Poisson–Gilbert graph. *Hint:* Use void space probabilities.   ◊

Hence, the only interesting question is about the global survival of the infection, which is only possible if the origin is connected to infinity in Φ_{D}. For this, we denote by λ_c the critical threshold for percolation of the Poisson-Boolean model with distance parameter R and recall the critical rate of infection

$$\lambda_{\mathrm{I,cr}}(x) = 2x - 1 - 2\sqrt{x^2 - x}$$

from Theorem 8.4.1. Note that the mapping $\lambda_{\mathrm{I,cr}} \colon [1, \infty) \to (0, 1]$, $x \mapsto \lambda_{\mathrm{I,cr}}(x)$ is strictly decreasing and that in the supercritical regime $\lambda_{\mathrm{D}} \geq \lambda_c$ in particular $\lambda_{\mathrm{D}}|B_R(o)| \geq 1$, see for example [Pen91, Equation 6.2]. We can now present the following result about global extinction of the infection from [CHJW19].

Theorem 8.4.3 (Global Extinction for Chase-Escape Models on Poisson–Gilbert Graphs)
If $0 \leq \lambda_{\mathrm{D}} < \lambda_c$, then the infection can not globally survive. Further, if $\lambda_{\mathrm{D}} \geq \lambda_c$ and $\lambda_{\mathrm{I}} \geq 0$, then there exists $\lambda_{\mathrm{W,cr}}(\lambda_{\mathrm{I}}, \lambda_{\mathrm{D}}) < \infty$ such that for all $\lambda_{\mathrm{W}} > \lambda_{\mathrm{W,cr}}(\lambda_{\mathrm{I}}, \lambda_{\mathrm{D}})$ the infection cannot globally survive. Finally, if $\lambda_{\mathrm{D}} \geq \lambda_c$ and $\lambda_{\mathrm{I}} \leq \lambda_{\mathrm{I,cr}}(\lambda_{\mathrm{W}}|B_R(o)|)$, then $\lambda_{\mathrm{W,cr}}(\lambda_{\mathrm{I}}, \lambda_{\mathrm{D}}) = 0$.

In words, global survival of the infection is impossible if either the process of devices is subcritical or the intensity of the white-knight process is sufficiently large. However, if the infection rate is too small, then any positive intensity of white knights is enough to exclude global survival. The first part of the proof is immediate, the second part goes along the lines of the proof of Theorem 8.4.1, whereas the third part uses discrete percolation arguments.

As a complement, from [CHJW19] we present the following statement about global survival in the supercritical percolation regime for the susceptible devices and sufficiently large infection rates.

Theorem 8.4.4 (Global Survival for Chase-Escape Models on Poisson–Gilbert Graphs)

For all $\lambda_D > \lambda_c$ and $\lambda_W \geq 0$ there exists $\lambda_{I,cr}(\lambda_D, \lambda_W) < \infty$ such that for all $\lambda_I > \lambda_{I,cr}(\lambda_D, \lambda_W)$, we have that the infection survives globally.

The proof again rests on a comparison with suitable percolation models in discrete space. Theorems 8.4.3 and 8.4.4 can be summarized in the phase diagram as sketched in ◻ Fig. 8.5. Let us note that in the chase-escape model on Poisson–Gilbert graphs there is a lack of monotonicities with respect to the parameters. This is partially already present in the model on $\mathbb{Z}^d$ and makes proofs substantially more difficult. In particular, uniqueness of the phase-separating line remains a hard open problem.

Remark 8.4.5 (Annealed Versus Quenched) The modeling of malware propagation via IPS on a random device-to-device network can be seen as part of the theory of IPS in random environment. More abstractly, the environment is a first layer of randomness, and the IPS

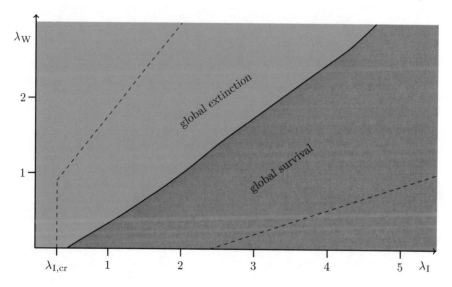

◻ **Fig. 8.5** Illustration of the phase diagram of global survival and extinction based on simulations from [CHJW19]. The black line indicates the anticipated critical curve of the infection rate versus the white-knight intensity for $d = 2, R = 1$ and $\lambda_D = 3$. The dashed lines indicate the regimes covered by Theorems 8.4.3 and 8.4.4

is the second layer, defined conditional on the first layer, the environment. Whenever two layers of randomness are present where the second-layer process is defined conditional on the first-layer process, there are, in principle, two related but different perspectives. First, the second-layer process can be observed in expectation over the first layer, this is the *annealed* setting. Second, the second-layer process can be considered almost-surely, i.e., for a fixed realization of the first layer, which is usually referred to as the *quenched* perspective.

Stochastic processes in random environment are the most common examples of such models, like the IPS on a random node set, as we considered here. A priori, in the quenched setting, the second-layer process depends on every detail of the environment, but in large-box approximations these many influences can often be replaced by some few deterministic quantities. More explicitly, the two perspectives are often connected via ergodic theorems, see ▶ Chap. 6, where in the appropriate ergodic limit, the annealed system turns out to be an almost-sure limit of the quenched system.

Note that this type of two-layer randomness already appeared multiple times in this monograph, for example in ▶ Sect. 2.4 about marked point processes or in ▶ Chap. 3 about Cox point processes. Usually the analysis starts with the consideration of the annealed setting, which is also what is featured in the book. However, lifting results towards the quenched perspective offers an interesting field for future research.

Bibliography

[And12]   W.J. ANDERSON, *Continuous-Time Markov Chains: An Applications-Oriented Approach*, Springer, New York, (2012).

[AsmGly07]   S. ASMUSSEN and P.W. GLYNN, *Stochastic Simulation: Algorithms and Analysis*, Springer, New York, (2007).

[BezGri90]   C. BEZUIDENHOUT and G. GRIMMETT, The critical contact process dies out, *Ann. Probab.* **18**, 1462–1482 (1990).

[BolRio06]   B. BOLLOBÁS and O. RIORDAN, *Percolation*, Cambridge University Press, New York (2006).

[BolRio08]   B. BOLLOBÁS and O. RIORDAN, Percolation on random Johnson–Mehl tessellations and related models, *Probab. Theory Relat. Fields* **140:3–4**, 319–343 (2008).

[Bil68]   P. BILLINGSLEY, *Convergence of Probability Measures*, Wiley, New York (1968).

[BacBla09a] F. BACCELLI and B. BŁASZCZYSZYN, *Stochastic Geometry and Wireless Networks, Volume I: Theory*, Now Publishers Inc. (2009).

[BacBla09b] F. BACCELLI and B. BŁASZCZYSZYN, *Stochastic Geometry and Wireless Networks, Volume II: Applications*, Now Publishers Inc. (2009).

[CHJ19]   E. CALI, C. HIRSCH and B. JAHNEL, Continuum percolation for Cox processes, *Stoch. Process. Their Appl.* **129:10**, 3941–3966 (2019).

[ChaHar14]   S. CHATTERJEE and M. HAREL, Localization in random geometric graphs with too many edges. *Ann. Probab.* **48:2**, 574–621 (2020).

[Cou12]   T. COURTAT, Promenade dans les cartes de villes-phénoménologie mathématique et physique de la ville-une approche géométrique, *Ph.D. dissertation, Université Paris-Diderot, Paris* (2012).

[CoxDur81]   J.T. COX and R. DURRETT, Some limit theorems for percolation processes with necessary and sufficient conditions, *Ann. Probab.* **9**, 583–603 (1981).

[CKMS13]   S.N. CHIU, W.S. KENDALL, J. MECKE and D. STOYAN, *Stochastic Geometry and Its Applications*, J. Wiley & Sons, Chichester (2013).

[DalVer03]   D.J. DALEY and D. VERE-JONES, *An Introduction to the Theory of Point Processes, Volume I: Elementary Theory and Methods*, Second Edition, Springer (2003).

[DalVer08]   D.J. DALEY and D. VERE-JONES, *An Introduction to the Theory of Point Processes, Volume II: General Theory and Structure*, Second Edition, Springer (2008).

[DemZei10]   A. DEMBO and O. ZEITOUNI, *Large Deviations Techniques and Applications*, Volume 38 of *Stochastic Modelling and Applied Probability*, Springer (2010).

[Der19]   D. DEREUDRE, *Stochastic Geometry: Introduction to the Theory of Gibbs Point Processes*, Springer (2019).

[DeuStr89]   J.-D. DEUSCHEL and D.W. STROOCK, *Large Deviations*, AMS Chelsea Publishing, Volume 342 (1989).

© Springer Nature Switzerland AG 2020
B. Jahnel, W. König, *Probabilistic Methods in Telecommunications*, Compact Textbooks in Mathematics,
https://doi.org/10.1007/978-3-030-36090-0

[DBT05]   O. DOUSSE, F. BACCELLI and P. THIRAN, Impact of interferences on connectivity in
            ad hoc networks, *IEEE/ACM Transactions on Networking (TON)* **13**, 425–436 (2005).

[DFMMT06]   O. DOUSSE, M. FRANCESCHETTI, N. MACRIS, R. MEESTER and P. THIRAN,
            Percolation in the signal to interference ratio graph, *J. Appl. Probab.* **43**, 552–562
            (2006).

[Dur88]   R. DURRETT, *Lecture Notes on Particle Systems and Percolation*, Brooks/Cole Pub Co
            (1988).

[DurGri82] R. DURRETT and D. GRIFFEATH, Contact processes in several dimensions, *Probab.
            Theory Relat. Fields* **49:4**, 535–552 (1982).

[DJT18]   R. DURRETT, M. JUNGE and S. TANG, Coexistence in chase-escape, *Electron.
            Commun. Probab.* **25:14** (2020).

[FraMee08] M. FRANCESCHETTI and R. MEESTER, *Random Networks for Communication: From
            Statistical Physics to Information Systems*, Cambridge University Press (2008).

[GanTor08] A.J. GANESH and G.L. TORRISI, Large deviations of the interference in a wireless
            communication model, *IEEE Trans. Inf. Theory* **54:8** (2008).

[Geo11]   H.-O. GEORGII, *Gibbs Measures and Phase Transitions*. De Gruyter Studies in
            Mathematics (2011).

[GeoZes93] H.-O. GEORGII and H. ZESSIN, Large Deviations and the maximum entropy principle
            for marked point random fields, *Probab. Theory Relat. Fields* **96**, 177–204 (1993).

[Gil61]   E.N. GILBERT, Random plane networks, *J. of the Society for Industrial & Appl. Math.*
            9, 533–543 (1961).

[Gra11]   R.M. GRAY, *Entropy and Information Theory*, Springer (2011).

[Gri89]   G.R. GRIMMETT, *Percolation*, Springer, New York (1989).

[HägPem00] O. HÄGGSTRÖM and R. PEMANTLE, Absence of mutual unbounded growth for almost
            all parameter values in the two-type Richardson model, *Stoch. Process. Their Appl.*
            90:2, 207–222 (2000).

[Hän12]   M. HÄNGGI, *Stochastic Geometry for Wireless Networks*, Cambridge University Press
            (2012).

[CHJW19]   E. CALI, A. HINSEN, B. JAHNEL and J.-P. WARY, Phase transitions for chase-escape
            models on Poisso–Gilbert graphs, *Electron. Commun. Probab.* **25:25** (2020).

[Hir16]   C. HIRSCH, Bounded-hop percolation and wireless communication, *J. Appl. Probab.*
            53:3, 833–845 (2016).

[HJKP18]   C. HIRSCH, B. JAHNEL, P. KEELER and R. PATTERSON, Large deviations in
            relay-augmented wireless networks, *Queueing Systems* **88**, 3–4 (2018).

[HJKP16]   C. HIRSCH, B. JAHNEL, P. KEELER and R. PATTERSON, Large-deviation principles for
            connectable receivers in wireless networks, *Adv. Appl. Probab.* **48**, 1061–1094 (2016).

[HJT19]   C. HIRSCH, B. JAHNEL and A. TÓBIÁS, Lower large deviations for geometric
            functionals, *arXiv preprint arXiv:1910.05993* (2019).

[JahTob19] B. JAHNEL and A. TÓBIÁS, Exponential moments for planar tessellations, *J. Stat.
            Phys.* **179**, 90–109 (2020).

[Jan86]   S. JANSON, Random coverings in several dimensions, *Acta Math.* **156**, 83–118 (1986).

[Kal97]   O. KALLENBERG, *Foundations of Modern Probability*, Probability and its
            Applications, Springer (1997).

[KatWei82] Y. KATZNELSON and B. WEISS, A simple proof of some ergodic theorems, *Israel J.
            Math.* **42:4**, 291–296 (1982).

[Kin95]   J. KINGMAN, *Poisson Processes*, Volume 3 of Oxford Studies in Probability, Oxford
            University Press, Oxford (1995).

[Kor05]   G. KORDZAKHIA, The escape model on a homogeneous tree, *Electron. J. Probab.* **10**,
            113–124 (2005).

[Kre85]   U. KRENGEL, *Ergodic Theorems*, De Gruyter, Berlin (1985).

[KTB13]   D.P. KROESE, T. TAIMRE and Z.I BOTEV, *Handbook of Monte Carlo Methods*, J. Wiley & Sons, New York (2013).

[LasPen17]   G. LAST and M. PENROSE, *Lectures on the Poisson Process*, IMS Textbook, Cambridge University Press (2017).

[vLie12]   M. VAN LIESHOUT, An introduction to planar random tessellation models, *Spat. Stat.* **E76**, 40–49 (2012).

[Lig85]   T.M. LIGGETT, *Interacting Particle Systems*, Springer, New York (1985).

[Lig10]   T.M. LIGGETT, *Continuous Time Markov Processes: An Introduction*, Volume 113 Graduate Studies in Mathematics, AMS (2010).

[Lig13]   T.M. LIGGETT, *Stochastic Interacting Systems: Contact, Voter and Exclusion Processes*, Volume 324 Springer, New York (2013).

[LauZuy08]   C. LAUTENSACK and S. ZUYEV, Random Laguerre tessellations, *Adv. in Appl. Probab.* **40:3**, 630–650 (2008).

[Mac03]   D.J.C. MACKAY, *Information Theory, Inference and Learning Algorithms*, Cambridge University Press (2003).

[Mat75]   G. MATHERON, *Random Sets and Integral Geometry*. John Wiley & Sons, London (1975).

[MKM78]   K. MATTHES, J. KERSTAN and J. MECKE, *Infinitely Divisible Point Processes*, John Wiley & Sons, Chichester (1978).

[MeeRoy96]   R. MEESTER and R. ROY, *Continuum Percolation*, Cambridge University Press, Cambridge (1996).

[MenSin16]   L. MÉNARD and A. SINGH, Percolation by cumulative merging and phase transition for the contact process on random graphs, *Ann. Sci. Éc. Norm. Supér. (4)* **49:5**, 1189–1238 (2016).

[Møl12]   J. MØLLER, *Lectures on Random Voronoi Tessellations*, Springer (2012).

[MølSto07]   J. MØLLER and D. STOYAN, *Stochastic Geometry and Random Tessellations*, Tech. rep.: Department of Mathematical Sciences, Aalborg University (2007).

[OBSC00]   A. OKABE, B. BOOTS, K. SUGIHARA and S.N. CHUI, *Spatial Tessellations: Concepts and Applications of Voronoi Diagrams*, J. Wiley & Sons, Chichester (2000).

[Pem92]   R. PEMANTLE, The contact process on trees, *Ann. Probab.* **20:4**, 2089–2116 (1992).

[Pen91]   M.D. PENROSE, On a continuum percolation model, *Adv. Appl. Probab.* **23:3**, 536–556 (1991).

[Pen03]   M.D. PENROSE, *Random Geometric Graphs*, Volume 5 of *Oxford Studies in Probability*, Oxford University Press, Oxford (2003).

[QuiZif07]   J.A. QUINTANILLA and R.M. ZIFF, Asymmetry in the percolation thresholds of fully penetrable disks with two different radii, *Phys. Rev.* **E76**, 051–115 (2007).

[RasSep15]   F. RASSOUL-AGHA and T. SEPPÄLÄINEN, *A Course on Large Deviation Theory with an Introduction to Gibbs measures*, Graduate Studies in Mathematics **162**, AMS (2015).

[Res87]   S.I. RESNICK, *Extreme Values, Regular Variation and Point Processes*, Applied Probability. A Series of the Applied Probability Trust 4. Springer, New York (1987).

[Ric73]   D. RICHARDSON, Random growth in a tessellation, *Math. Proc. Camb. Philos. Soc.* **74:3**, 515–528 (1973).

[SepYuk01]   T. SEPPÄLÄINEN and J. E. YUKICH, Large deviation principles for Euclidean functionals and other nearly additive processes, *Probab. Theory Relat. Fields*, **120:3**, 309–345 (2001).

[Sha48]   C.E. SHANNON, A Mathematical theory of communication, *Bell Syst. Tech. J.* **27**, 379–423 (1948).

[Tob18]   A. TÓBIÁS, Signal to interference ratio percolation for Cox point processes, *arXiv preprint arXiv:1808.09857* (2018).

[Val10]   D. VALESIN, Multitype contact process on $\mathbb{Z}$: extinction and interface, *Electron. J. Probab.* **15**, 2220–2260 (2010).

[Wie39]   N. WIENER, The ergodic theorem, *Duke Math. J.* **5:1**, 1–18 (1939).

[YCG11]   C. YAO, G. CHEN and T.D. GUO, Large deviations for the graph distance in supercritical continuum percolation, *J. Appl. Probab.* **48**, 154–172 (2011).

Index

© Springer Nature Switzerland AG 2020
B. Jahnel, W. König, *Probabilistic Methods in Telecommunications*, Compact Textbooks in Mathematics,
https://doi.org/10.1007/978-3-030-36090-0

Printed in the United States
By Bookmasters